DATE DUE

Magical Mathematics

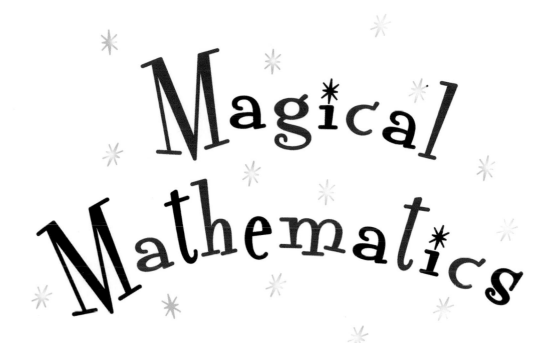

Magical Mathematics

THE MATHEMATICAL IDEAS THAT ANIMATE GREAT MAGIC TRICKS

PERSI DIACONIS
AND
RON GRAHAM

With a foreword by Martin Gardner

PRINCETON UNIVERSITY PRESS
PRINCETON AND OXFORD

Copyright © 2012 by Princeton University Press

Published by Princeton University Press, 41 William Street,
Princeton, New Jersey 08540
In the United Kingdom: Princeton University Press, 6 Oxford Street,
Woodstock, Oxfordshire OX20 1TW
press.princeton.edu

Library of Congress Cataloging-in-Publication Data

Diaconis, Persi.
    Magical mathematics : the mathematical ideas that animate great magic tricks /
Persi Diaconis, Ron Graham ; with a foreword by Martin Gardner.
        p. cm.
    Includes bibliographical references and index.
    ISBN 978-0-691-15164-9 (hardback)
    1. Card tricks—Mathematics. I. Graham, Ron, 1950– II. Title.
    GV1549.D53 2011
    793.8'5—dc23

                                                                    2011014755

British Library Cataloging-in-Publication Data is available

This book has been composed in ITC New Baskerville
Printed on acid-free paper. ∞
Printed in the United States of America

10 9 8 7 6 5 4 3 2

To our wives,

SUSAN AND FAN

CONTENTS

Foreword   ix

Preface   xi

1   MATHEMATICS IN THE AIR   1
Royal Hummer   8
Back to Magic   15

2   IN CYCLES   17
The Magic of de Bruijn Sequences   18
Going Further   25

3   IS THIS STUFF ACTUALLY GOOD FOR ANYTHING?   30
Robotic Vision   30
Making Codes   34
To the Core of Our Being   38
This de Bruijn Stuff Is Cool but Can It Get You a Job?   42

4   UNIVERSAL CYCLES   47
Order Matters   47
A Mind-reading Effect   52
Universal Cycles Again   55

5   FROM THE GILBREATH PRINCIPLE TO THE MANDELBROT SET   61
The Gilbreath Principle   61
The Mandelbrot Set   72

6   NEAT SHUFFLES   84
A Mind-reading Computer   85
A Look Inside Perfect Shuffles   92

A Look Inside Monge and Milk Shuffles   96
A Look Inside Down-and-Under Shuffles   98
All the Shuffles Are Related   99

7   THE OLDEST MATHEMATICAL ENTERTAINMENT?   103
The Miracle Divination   105
How Many Magic Tricks Are There?   114

8   MAGIC IN THE *BOOK OF CHANGES*   119
Introduction to the *Book of Changes*   121
Using the *I Ching* for Divination   122
Probability and the *Book of Changes*   125
Some Magic (Tricks)   127
Probability and the *I Ching*   136

9   WHAT GOES UP MUST COME DOWN   137
Writing It Down   138
Getting Started in Juggling   145

10 STARS OF MATHEMATICAL MAGIC (AND
SOME OF THE BEST TRICKS IN THE BOOK)   153
Alex Elmsley   156
Bob Neale   160
Henry Christ   173
Stewart James   181
Charles Thornton Jordan   189
Bob Hummer   201
Martin Gardner   211

11 GOING FURTHER   220

12 ON SECRETS   225

Notes   231

Index   239

If you are not familiar with the strange, semisecret world of modern conjuring you may be surprised to know that there are thousands of entertaining tricks with cards, dice, coins, and other objects that require no sleight of hand. They work because they are based on mathematical principles.

Consider, for example, what mathematicians call the Gilbreath Principle, named after Norman Gilbreath, its magician discoverer. Arrange a deck so the colors alternate, red, black, red, black, and so on. Deal the cards to form a pile about equal to half the deck, then riffle shuffle the piles together. You'll be amazed to find that every pair of cards, taken from the top of the shuffled deck, consists of a red card and a black card! Dozens of beautiful card tricks—the best are explained in this marvelous book—exploit the Gilbreath Principle and its generalizations.

Although you can astound friends with tricks based on this principle, they are in this book for another reason. The principle turned out to have applications far beyond trivial math. For example, it is closely related to the famous Mandelbrot set, an infinite fractal pattern generated on a computer screen by a simple formula.

But that is not all. The Dutch mathematician N. G. de Bruijn discovered that the Gilbreath principle applies to the theory of Penrose tiles (two shapes that tile the plane only in a nonperiodic way) as well as to the solid form of Penrose tiles, which underlies what are called quasicrystals. Still another application of the principle, carefully explained in this book, is to the design of computer algorithms for sorting procedures.

The authors are eminent mathematicians. Ron Graham, retired from Bell Labs and now a professor at the University of California, San Diego, is an expert on combinatorial math. Persi Diaconis is an equally famous statistician at Stanford University. Each man has a hobby. Ron is a top juggler. Persi is a skilled card magician.

You will learn from their book the math properties of unusual shuffles: the faro, the milk shuffle, the Monge shuffle, and the Australian or down-and-under shuffle. You will learn some tricks using the *I Ching*, an ancient Chinese fortune-telling volume. You will learn how parity (odd or even) can play a roll in magic as well as provide powerful shortcut proofs.

Not only is this book a superb, informally written introduction to mathematical magic, but near the book's end the authors supply pictures and biographical sketches of magicians who have made the greatest contributions to mathematical magic, from the reclusive Charles Jordan to the eccentric Bob Hummer.

Best of all, you will be introduced to many little-known theorems of advanced mathematics. The authors lead you from delightful self-working magic tricks to serious math, then back again to magic. It will be a long time before another book so clearly and entertainingly surveys the vast field of mathematical hocus-pocus.

Martin Gardner
Norman, OK
April, 2010

The two of us have been mixing entertainment with mathematics for most of our lives. We started off on the entertainment side, one as a magician, the other as a juggler and trampolinist. We were seduced into studying mathematics by . . . well, the stories that are told in this book. Both of us now make a living doing mathematics; teaching, proving, and conjecturing.

The two fields have been shuffled together for us, with frequently performed talks on mathematics and magic tricks and the mathematics of juggling. The connections go deeper. Some magic tricks use "real mathematics" and lead to questions beyond the limits of modern mathematics (see our chapter on shuffling cards). Sometimes, we have been able to solve the math problems and create new magic tricks (see chapter 2).

Both of our worlds have a dense social structure; thousands of players turning ideas over and over. Some of this wisdom of the ages is woven through our book. In addition to hundreds of friends and colleagues, dozens of people have made sustained contributions to this book.

On the magic front, Steve Freeman, Ricky Jay, Bob Neale, and Ronald Wohl have been coworkers, selflessly contributing their brilliance and wisdom. The students in our Magic and Mathematics classes at Harvard and Stanford have all helped. We particularly thank Joe Fendel. We had the benefit of amazing, insightful readings of our text by Art Benjamin, Steve Butler, Colm Mulcahy, and Barry Mazur. Their combined comments rivaled the length of our book. Laurie Beckett, Michael Christ, Jerry Ferrell, Albrecht Heeffer, Bill Kalush, Mitsunobu Matsuyama, and Sherry Wood went out of their way to help us out. Our editors at Princeton University Press, Ed Tenner, Vickie Kearn, and Mark Bellis have been crucial allies.

Our families, Fan Chung Graham, Ché Graham, and Susan Holmes, have helped in so many ways that we can't find a number system rich enough to list them. Fan's mathematical work appears in chapters 2–4, and Ché and Susan shot (and reshot) numerous photos. Susan also contributed to the history and many other chapters.

We hope that our book will shine a friendly light on the corners of the world that are our homes.

Thanks and welcome.

Persi Diaconis and Ron Graham

Magical Mathematics

Chapter 1

MATHEMATICS IN THE AIR

Most mathematical tricks make for poor magic and in fact have very little mathematics in them. The phrase "mathematical card trick" conjures up visions of endless dealing into piles and audience members wondering how long they will have to sit politely. Our charge is to present entertaining tricks that are easy to perform and yet have interesting mathematics inside them. We cannot do this without your

Figure 1. Four cards

Figure 2. Four cards in a packet

Figure 3. Looking at bottom card

help. To get started, please go find four playing cards. They can be any four cards, all different or the four aces. It doesn't matter. Let us begin by performing the trick for you. Since we can do it without being present, you'll be able to do it for a friend on the phone. After practicing, try calling your kid brother or your mom and perform the following.

Have a look at the bottom card of the packet. That's your card and you have to remember it.

Next, the cards are going to be mixed by some simple instructions. Put the top card on the bottom of the packet. Turn the current top card face-up and place it back on top.

Figure 4. Top card placed on bottom

Figure 5. Current top card turned face-up

Now, give the packet a cut. It doesn't matter how many cards you cut from top to bottom: one, two, three, or four (which is the same as none). Next, spread off the top two cards, keeping them together, and turn them over, placing them back on top.

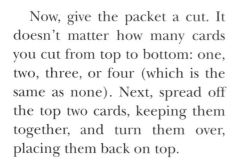

Figure 6. Cutting the deck

Figure 7. Spreading off and turning over the top two

Cut the cards at random again and then turn the top two over. Give them another cut and turn two over.

Give them a final cut. This cutting and turning has mixed the cards in a random fashion. There is no way anyone can know the order. Remember the name of your card! We're going to find it together.

Figure 8. Cutting again

Figure 9. Turning over two again

Figure 10. Another cut

Figure 11. Turning over two

Figure 12. Turning over the top card

Figure 13. Putting it on the bottom

Figure 14. Putting the top card on the bottom

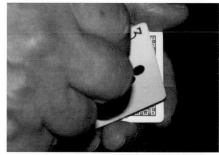

Figure 15. Turning over the top card

Figure 16. The "oddball" card

Figure 17. The chosen card

Turn the top card over (if it's face-down, turn it face-up; if it's face-up, turn it face-down). Put this card on the bottom of the packet.

Put the current top card on the bottom of the packet without turning it over. Finally, turn the top card over and place it back on top.

Now, we're done. Name your card. Spread out the packet of four. You'll find three cards facing one way and your card facing the opposite way!

When we perform this trick with a live audience in the same room, we try to work it on a man with a tie or a woman with a scarf. We give him or her the four cards with instructions to shuffle, peek at the bottom card, and follow the instructions above until he or she has cut and turned over two a few times. We then ask our subject to put the four cards behind his or her back. The rest of the instructions are carried out with the cards concealed this way. When the cutting and turning phase is finished, we stare intently at the person's midsection in giving the final two steps of instructions as if we were looking through our subject. Before the final line of instruction we reach over and move the tie or scarf as if it were blocking our view. We have him or her name the card before bringing out the packet.

We have used this trick for an audience of a hundred high school students—each student received a packet of four cards, and the trick was worked simultaneously for all of them. It's a charming trick and really seems to surprise people.

Okay. How does it work? Let's start by making that your problem: How does it work? You'll find it curiously difficult to give a clear explanation. In twenty years of teaching, asking students to try to explain this trick, we have yet to have anyone give a truly clear story. The plan is to lead you through this in stages (it has some math in it). The solution comes later in this chapter. Before proceeding, let's generalize.

The trick is known as Baby Hummer in magic circles. It was invented by magician Charles Hudson as a variation on an original trick by a truly eccentric genius named Bob Hummer. We'll learn a lot more about Hummer as we go along. Here is his original use of the principle we're trying to explain.

Take any ten cards. Have them all face-down and hold them as if you were about to deal in a card game. Go through the following procedure, which mixes the cards face-up and face-down: Spread the top two cards off and turn them over, placing them back on top. Give the cards a straight cut (see figure 6). Repeat this "turn two and cut at random" procedure as often as you like. The cards will be in an unpredictable mess. To find the order in the mess, proceed as follows: Go through the packet, reversing every second card (the cards in positions 2, 4, 6, 8, and 10). You will find exactly five cards face-up, no matter how many times the "turn two and cut at random" procedure was repeated.

Hummer marketed this trick in a privately printed manuscript called "Face-up/Face-down Mysteries" (1942).[1] This ten-card trick does not play as well for audiences as the Baby Hummer we started with. Hummer introduced a kind of swindle as a second phase. After showing that five cards are face-up and five cards are face-down, casually rearrange the cards so that the face-up and face-down cards alternate up, down, up, down, and so on. Hand the ten cards to a spectator who is instructed to put the cards under the table (or behind his or

Figure 18. Reversing every second card

her back). Have the spectator repeat the "turn two and cut at random" procedure a few times. Take the cards back without looking at them. Now, with the cards under the table (or behind your back), remove every second card as before and turn them over. You will find that the cards all face the same way.

Again, one may ask, why does this work? *Just what properties of the arrangement are preserved by Hummer's "turn two and cut at random" procedure?* To think about Hummer's "turn two and cut at random" mixing scheme, we find it helpful to have a way of writing down all the possible arrangements that can occur. Instead of working with a deck of four or ten cards, one can just as easily work with a general deck of even size. We work with $2n$ cards (so, if $n = 2$ then $2n = 4$, or if $n = 5$ then $2n = 10$). As will be seen in a while, decks of odd size are a different kettle of fish. We can indicate the exact arrangement of $2n$ cards, some face-up and some face-down, by writing the numbers on the cards in order and identifying face-up with a bar on top of a number. Thus, a four-card deck with a face-up 3 on top, a face-down 1 next, a face-down 4 next, and a face-up 2 at the bottom is denoted $\overline{3}, 1, 4, \overline{2}$. For a deck of ten cards, a possible arrangement is $2, \overline{1}, \overline{4}, 8, 6, \overline{5}, \overline{3}, \overline{10}, \overline{7}, 9$.

The symbols $1, 2, 3, \ldots, 2n$ can be arranged in $1 \times 2 \times 3 \times 4 \times \cdots \times 2n$ ways. This number is often denoted as $(2n)!$ (read "$2n$ factorial"). Each such arrangement can be decorated with bars in $2 \times 2 \times 2 \times \cdots \times 2 = 2^{2n}$ ways (each of the $2n$ symbols can be barred or not). In all, this makes for $2^{2n} \times (2n)!$ distinct arrangements. This is a huge number even for a moderate n. For $2n = 4$, it is $2^4 \times 4! = 16 \times 24 = 384$. For $2n = 10$, it is $3,715,391,200$ (close to four billion). This is the maximum possible number of arrangements. As we will see, not all of these are achievable if we start with a face-down deck using Hummer's "turn two and cut at random" process.

Before we give the general answer, here is a starter result that shows that many of the $2^{2n} \times (2n)!$ arrangements are not achievable. This result also clearly explains why Hummer's ten-card trick works. We present it as a simple theorem to show that theorems can grow anywhere.

> THEOREM. Let a deck of $2n$ cards start all face-down. After any number of "turn two and cut at random" operations, the following regularity is forced:
>
> The number of face-up cards at *even* positions
> equals
> the number of face-up cards at *odd* positions.

Normally, we will put our proofs at the end of each chapter. However, we give the proof for this here. What we want to prove is certainly true when we start—there are no face-up cards in either even or odd positions at the start. Suppose the statement of the theorem holds after some fixed number of shuffles. Observe that it still holds after a single card is cut from top to bottom. Therefore, it holds if any number of cards is cut from top to bottom. So the result to be proved holds for any number of cuts. Finally, suppose that the result to be proved holds for the current deck. Note that the current deck may well have cards face-up and face-down. Let us argue that it continues to hold after the top two cards are turned over and put back on top. We see this by considering all possible arrangements of the top two cards. They may be:

down, down   down, up   up, down   up, up.

After turning two, these four possibilities become:

up, up   down, up   up, down   down, down.

In the middle two cases, the up-down pattern hasn't changed, so the statement holds after turning two if it held at the start. In the first case, the odd positions and the even positions each have one more up card. Since the numbers of face-ups in even and odd positions were equal before we turned two, they are equal after. The same argument works in the last case. This covers all cases and proves the theorem.

From the theorem, it is a short step to see why Hummer's trick works. Start with $2n$ cards face-down ($2n = 10$ for Hummer). After any number of "turn two and cut at random" shuffles, there will be some number of face-up cards. Let A be the number of face-up cards among the n cards at even positions. There must be $n - A$ face-down cards among the even positions since there are n cards in even positions. By the theorem, the same holds for the n cards at odd positions—A face-up and $n - A$ face-down. If you remove the cards at odd positions and turn them over, this gives $n - A$ face-up cards to add to the A face-up cards at even positions. This makes $(n - A) + A = n$ face-up cards in all. Of course, the other n cards are face-down. The conclusion is forced.

Did the proof we just gave ruin the trick? For us, it is a beam of light illuminating a fuzzy mystery. It makes us just as happy to see clearly as to be fooled.

To check your understanding, we mention that, in magic circles, Hummer's principle is sometimes called CATO for "cut and turn over two." This is in opposite order to the "turn over two and cut." The theorem holds for CATO as well as "cut and turn over four" or "turn over an even number and cut."

Later in this chapter we show that exactly $2 \times (2n)!$ arrangements are achievable and just which ones these are. This more general result implies the theorem we just proved and, indeed, all possible theorems about Hummer's mixing process.

In the meantime we turn to the question: How can a really good trick be twisted out of this math? We give as an answer a closely guarded secret of one of the great card men of the present era. Steve Freeman has given us permission to explain what we think is an amazing amplification of Hummer's shuffles. We explain it by first describing the effect and then the modus operandi. Those wishing to understand why it works will have to study the math at the end of the chapter.

ROYAL HUMMER

First, the effect as the audience sees it. The performer hands the spectator about one-third of the deck, asking that the cards be thoroughly shuffled. Taking the cards back from the spectator, the performer explains that the cards will be further mixed, face-up and face-down, at the spectator's discretion, to make a real mess. The cards are dealt off in pairs, the spectator deciding each time if they should be left as is or turned over. This is repeated with the cards in groups of four. At this point, there is a pile of face-up/face-down cards on the table. The performer says, "I think you must agree that the cards are truly randomly distributed." The spectator gets one more decision—after the performer deals the cards into two piles (left, right, left, right, and so on) the spectator chooses a pile, turns it over, and puts it on top of the other pile. For the denouement, the performer explains that the highest hand in poker, the perfect poker hand, is a royal flush—ace, king, queen, jack, and ten, all of the same suit. The cards are spread and there are exactly five face-down cards. "Five cards—that just makes a poker hand." The five are turned over one at a time—they form a royal flush.

That's the way the trick looks. Here is how it works. Before you begin, look through the deck of cards, as if checking to see if the deck is complete, and place one of the royal flushes on top (they do not have to be in order). Remove the top twenty or so cards. The exact number doesn't matter as long as it's even and contains the royal flush. Have the spectator shuffle these cards. Take the cards back, turn them all face-up, and start spreading through them as you explain the next phase. Look at the first two cards.

Figure 19. Neither card is in the flush

Figure 20. Turning the second card

1. If neither one is in the royal flush, leave the first card face-up and flip the second card face-down, keeping both in their original position (you may use the first card to flip over the second one).

Figure 21. Second card is turned over

Figure 22. Pair is placed on the table

Figure 23. Only the first card is in the flush

Figure 24. First card is turned over

2. If the first one is in the royal flush and the second one is not, flip the first one face-down and then flip the second one face-down (they stay in their original positions).

Figure 25. Second card is also turned over, with the cards kept in order

Figure 26. Pair is placed on the table

3. If the second one is in the royal flush and the first one is not, leave both face-up.

Figure 27. Only the second card is in the flush

Figure 28. Placing the pair face-up in order on the table

4. If both are in the royal flush, flip the first one face-down and leave the second one face-up.

Figure 29. Both cards are in the flush

Figure 30. Turning the first card over

Figure 31. Placing the pair on the table

Figure 32. The completed arrangement

The pairs may be dropped onto the table in a pile after each is adjusted or passed into the other hand. Work through the packet a pair at a time, using the same procedure for each pair. If, by chance, you wind up with an odd number of cards, add an extra card from the rest of the deck.

Now, take off the cards in pairs, asking the spectator to decide, for each pair, whether to "leave them or turn them," and put them into a pile on the table as dictated. When done, you can pick up the pile and go through the "leave them or turn them" process for pairs as before (or in sets of four, if desired). To finish, deal the cards into two piles (left, right, left, right, . . .). Have the spectator pick up either pile, turn it over, and place it on the other one. If the royal flush cards are not facing down, turn the whole packet over before spreading.

This is a wonderful trick. It really seems as if the mixing is haphazard. The ending shocks people. It does take some practice but it's worth it—a self-working trick done with a borrowed deck (which doesn't have to be complete).

Perhaps the most important lesson to be learned is how a simple mathematical principle, introduced via a fairly weak trick, can be built into something special. This is the result of fifty years of sustained development by the magic community. People from all walks of life spent time turning the trick over, suggesting variations, and being honest about their success or failure. At the beginning and end were two brilliant contributors—Bob Hummer and Steve Freeman. We are in their debt.

A word about practice. The first times you run through, following the procedures (1)–(4) above, will be awkward and slow. After a hundred or so practice runs, you should be able to do it almost subconsciously, without really looking at the cards. A skillful performer must be able to patter along ("We will be turning cards face-up and face-down as we go. You will decide which is which . . ."). The whole proceeding must have a casual, unstudied feel to it. All of this takes practice.

In the rest of this chapter, we explain some math. As a warmup, let us argue that the Baby Hummer trick that begins this chapter always works. To begin with, in the original setup we have three cards facing one way and one card (which we'll call the "oddball") facing the other way. We'll say that cards in positions one and three (from the top) are "mates," as are cards in positions two and four. The setup instructions then force the chosen card and the oddball to be mates. It is easy to check that any "turn two and cut randomly" shuffle (or Hummer shuffle, for short) will preserve this relationship (there are basically only two cases to check). Finally, the finishing instructions have the effect of turning over exactly one card and its mate. This has the effect of forcing the chosen card to be the oddball. Again, two cases to check. End of story.

With all the variations, it is natural to ask just what can be achieved from a face-down packet, originally arranged in order 1, 2, 3, ... , $2n$, after an arbitrary number of Hummer shuffles. The following theorem delineates exactly what can happen.

THEOREM. After any number of Hummer shuffles of $2n$ cards, any arrangement of values is possible. However, the face-up/face-down pattern is constrained as follows: Consider the card at position i. Add one to its value if face-up. Add this to i. This sum is simultaneously even (or odd) for all positions i.

EXAMPLE. Consider a four-card deck in the final arrangement: $4, \overline{2}, \overline{1}, 3$. In position 1, the sum "position + value + (1 if face-up, and 0 if face-down)" is $1 + 4 + 0 = 5$, which is odd. The other three positions give

$$2 + 2 + 1 = 5, \; 3 + 1 + 1 = 5, \; 4 + 3 + 0 = 7,$$

all odd values.

REMARKS. The constraint in the theorem is the only constraint. All arrangements arrived at by the Hummer shuffling are bound by it, and any pattern of cards that satisfies the constraint is achievable by Hummer shuffles. An interesting unsolved problem is to figure out the minimum number of Hummer shuffles it takes to achieve any particular pattern.

Any property of Hummer shuffles is derivable from the theorem. We record some of these as corollaries.

COROLLARY 1. The number of achievable arrangements for a deck of $2n$ cards after Hummer shuffling is $2 \times (2n)!$.

REMARK. In mathematical language, the set of all achievable arrangements of $2n$ cards after Hummer shuffling forms a group.

COROLLARY 2. (Explanation of Hummer's original trick.) After any number of Hummer shuffles, the number of face-up cards at even positions equals the number of face-up cards at odd positions. Thus, if the even cards are removed and turned over, the total number of face-up cards is n.

PROOF: Consider the cards at even positions. If there are j even values, all of these must face the same way. Similarly, the $n - j$ odd values must face the other way. At the odd positions, there will be $n - j$ even values all facing in the opposite way to the even values at even positions. When the cards at even positions are removed and turned over, there are $j + (n - j) = n$ facing the same way, with the remaining n facing the opposite way.

COROLLARY TO COROLLARY 2. The argument underlying corollary 2 shows that in fact, after any number of Hummer shuffles followed with every other card removed and reversed, the cards originally at even positions all face the same way (likewise, the cards originally at odd positions all face the opposite way). Let us make this into a trick: Take five red cards and five black cards and arrange them in alternate colors in a face-down pile. Hummer shuffle any number of times, remove every other card, and reverse these. All the red cards face one way and all the black cards face the opposite way. This makes for quite a surprising trick. It may be endlessly varied. For example, remove four aces and six other cards. Place the aces in every second position (i.e., in positions two, four, six, and eight). Turn the bottom card face-up. The cards may be Hummer shuffled any number of times. Follow this by reversing every other card. The four aces will face opposite the remaining cards. Charles Hudson derived a number of entertaining tricks built on this idea. His Baby Hummer trick is explained above. Steve Freeman's Royal Hummer trick may be the ultimate version.

FINAL NOTES. We are not done understanding Hummer shuffles. The following two notes record a natural question (does it only work with even-sized decks?) and a new trick that comes from the analysis. There is a lot we still don't know. (For example, what about turning up three?)

NOTE 1. It is natural to wonder if the trick will work with an odd number of cards. It would be nice to ask the spectator to remove a random poker hand of five cards and begin the trick from here. We assume below that "turn two and cut at random" is used throughout.

There is one regularity: There will always be an even number of cards face-up. Alas, this is the only regularity. All $2^{n-1} \times n!$ signed arrangements of n cards (with n odd) are achievable with a deck of n cards.

Let us record one proof of this. First, any three cards can be manipulated so: $123 \to \overline{21}3 \to \overline{23}1 \to 321$, thus transposing positions 1 and 3. By doing this, any permutation of the even positions and also any permutation of the odd positions is possible. Consider transposing positions 1 and 3, and then 3 and 5, and then 5 and 7, . . . , and then $n-2$ and n. This results in 3, 2, 5, 4, 7, 6, . . . , $n-1$, 1. For example, with seven cards we get 3, 2, 5, 4, 7, 6, 1. Now transpose consecutive pairs in the even positions, moving the card labeled 2 to the right. This results in 3, 4, 5, . . . , $n-1$, 2, 1. Finally, cut the bottom two cards to the top. This all results in a simple transposition. As usual, this allows us to transpose any two consecutive cards and so finally to achieve any permutation of the labels.

Next, we show how to achieve any face-up/face-down pattern with an even number of face-up cards (where we use 0 to denote a face-down card, and 1 to denote a face-up card). This is achieved "two at a time." The following moves show how this can be done: $000 \ldots 0 \to 110 \ldots 0 \to 11110 \ldots 0 \to 10010 \ldots 0 \to 1001110 \ldots 0 \to 1000010 \ldots 0 \ldots$. After cutting, this gives any possible separation of the 1's (since n is odd). This shows that any pair can be turned face-up. Working one pair at a time shows that any pattern of an even number of cards can be turned face-up. Finally, combining our ability to create arbitrary arrangements of values with an arbitrary face-up/face-down pattern gives the final result.

From the above we may conclude that there is no real extension of Hummer's trick to an odd-sized packet. Of course, the two types of parity delineated above may form the basis for tricks.

NOTE 2. One reason for developing all this theory is the hope of inventing a new trick. Following is one that comes from our analysis.

Here is the effect. Ask a spectator to remove the ace through ten of spades and arrange them in order (ace–ten or ten–ace—it doesn't matter which). Then turn your back and have the spectator Hummer shuffle the ten-card packet any number of times. You can promise that you don't know anything about the order of the cards. Ask a spectator to name the values one at a time (from the top down) and you tell them if the cards are face-up or not.

From what was developed above, the only mystery is knowing the orientation (face-up or face-down) of the top card (all else follows). You simply guess! If correct, keep going. If wrong, rub your eyes and ask the spectator to concentrate. Try again! The trick as described may be done on the phone. Note you only need to know the odd/even values of consecutive cards to know their orientation.

Let us be the first to admit that, as described, this is a pretty poor trick. We hope that someone someplace will turn it over and around and come up with something performable. Please let us know (we'll shout it from the rooftops or, if you like, keep it as secret as secret can be).

BACK TO MAGIC

To conclude on a high note, here is Steve Freeman's favorite method of getting set for his Royal Hummer trick. This is a replacement for procedures (1)–(4) above. To begin, you have a packet of twenty or so cards that contains a royal flush, with all cards facing the same way. The royal flush is scattered throughout the packet. The cards will be split into two, one face-up packet in each hand. The hands alternately deal into one pile on the table, turning some cards over. At the end, the indifferent cards at even positions will be face-up. Indifferent cards at odd positions will be face-down. The royal flush cards are opposite. When the cards are dealt into two piles and one pile is turned over on the others, all of the indifferent cards face the same way and all of the royal flush cards are opposite.

To get comfortable with this, try a simple exercise: Take two packets of face-up cards, hold one in each hand in dealing position, and deal alternately into one pile, face-up on the table, left, right, left, right, etc. Do this until you can do it easily. Now, with the same start, try turning the left hand's cards face-down as they are dealt, so that the cards are placed down, up, down, up, and so forth. If this is awkward, try also turning the right hand's cards down (with the left's face-up) and then both hands' cards face-down. It is useful to keep the left/right alternation standardized throughout.

Now for the real thing. Begin with an even number of cards, less than half the deck, containing a royal flush, with all cards face-up. Split these into two roughly equal packets, held face-up in each hand. Each time, deal first from the left then from the right into the pile on the table. Observe the following rules:

1. If two indifferent cards show, deal the left face-up followed by the right face-down.
2. If two royal flush cards show, deal the left face-down followed by the right face-up.
3. With a flush card left and an indifferent card right, deal the left face-down followed by the right face-down.
4. With an indifferent card left and a flush card right, deal the left face-up and the right face-up.

If one hand runs out of cards, just split the remaining cards into two packets and continue. The trick continues as described above. Again, this takes practice to do naturally, accurately, and casually. Several dozen run-throughs might suffice.

Chapter 2

IN CYCLES

In this and the following two chapters, we explain a wonderful magic trick that leads to, and profits from, beautiful mathematics. The trick is one we have performed for drunks in seedy nightclubs, at Hubert's Flea Museum, and at a banquet of the American Mathematical Society. The trick really fools magicians, mathematicians, and "normal" people too. The mathematics involved begins with basic graph theory. Indeed, it uses ideas that started the subject of graph theory. It also needs tools of finite fields and combinatorics. At the heart of the trick are de Bruijn sequences. These are used in applications far beyond card tricks—for rhyming patterns in East Indian music, for robotic vision, and for making secret codes. The magical applications suggest variations that we call *universal cycles*. They need new mathematics, much of which doesn't yet exist (or, at least, is currently unknown).

The story is long enough that we tell it in three chapters. This chapter explains the trick and a bit about how it works. We explain what de Bruijn sequences are, show that they exist, and tell how to construct and count them. At the end, we give practical details on performing the trick.

Chapter 3 tells some stories: real-world applications of breaking and entering, industrial espionage, and decoding DNA, in which de Bruijn sequences are used. Chapter 4 describes some new magic tricks that involve generalizations of de Bruijn sequences. Understanding, constructing, and counting these new universal cycles leads us to the edge of what we know in mathematics. The chapters are self-contained but it all starts with the following magic trick.

THE MAGIC OF DE BRUIJN SEQUENCES

THE EFFECT

Here is what the audience sees: The performer has a deck of cards in its case (a few rubber bands around the deck will help ensure no disaster happens). The deck is tossed to an audience member who tosses it to another, and so on, until the deck is far at the back of the room. To actually perform this trick, you need at least five people in the audience but it is effective with an audience of a thousand. The final deck holder is asked to remove cards from the case, drop the case on the floor, and then give the deck a straight cut at a random position. The deck is passed to a second spectator who is asked to cut and pass it on. Finally, when a fifth spectator cuts, ask that the top card be taken off. The deck is then passed back to the fourth spectator who removes the current top card. Each of the five spectators in turn removes a card. The performer now asks, "This may sound strange but would each of you please look at your card, make a mental picture, and try to send it to me telepathically?" As this is done the performer concentrates and appears confused: "You're doing a great job, but there is too much information coming in for me to make sense of. Would all of you who have a red card please stand up and concentrate?" Suppose that the first and third spectators stand. The performer appears relieved and says, "That's perfect. I see a seven of hearts?" (One of the spectators shows that this is indeed the thought-of card.) "And a jack of diamonds? Yes." Now, focusing on the other three spectators, the performer names all three black cards.

There is nothing left out of the above description; the cards are well out of the performer's control and aren't tricked or marked in any way. So how does it work?

THE SECRET

The secret lies in the performer's innocent question: "Would all of you who have a red card please stand up and concentrate?" This question can be answered in thirty-two different ways: No one stands, only the first person, only the second person, only the first two, and so on, finishing with the possibility that all five spectators stand. With five spectators, each of whom stands or not, this makes $2 \times 2 \times 2 \times 2 \times 2 =$

32 possible answers. It just so happens that the deck handed out has thirty-two cards. (Sorry for leaving out that detail; the spectators never complain about it!) Of course, the deck is carefully arranged so each consecutive set of five cards has a unique color pattern.

To see the idea in simple form, suppose only three spectators were asked to remove cards. They can answer in $2 \times 2 \times 2 = 8$ ways, so an eight-card deck can be used. The eight possible answers are:

RRR RRB RBR RBB BRR BRB BBR BBB.

We want to find a sequence of eight R/B colors so that each consecutive set of three occurs just once. The reader can check that the sequence RRRBBBRB does the job. The first three use up RRR. The next three RRB. Further consecutive triples (going around the corner at the end) use up RBB, BBB, BBR, BRB, RBR, BRR. This is all eight used once and only once. Thus, this trick could be worked using the following eight cards: AH, 5D, 6H, 2S, 5S, KC, 7H, 8S, where H, D, C, S stand for hearts, diamonds, clubs, and spades, respectively. Of course, the values are irrelevant but the audience doesn't know this.

Before explaining the thirty-two-card version (and versions for larger decks), let us restate slightly. Replace the symbols R/B with the mathematicians' favorite: 1 and 0. Then RRRBBBRB becomes 11100010. A de Bruijn sequence with window length k is a zero/one sequence of length 2^k (this is just $2 \times 2 \times \cdots \times 2$, k times) such that every k consecutive digits appears just once (going around the corner). Thus, 11100010 is a de Bruijn sequence of window length 3. If we have a de Bruijn sequence of window length k, we can perform the trick with 2^k cards. As will emerge in the next chapter, de Bruijn sequences with large values of k are needed in applications. We need one with $k = 5$. The puzzle-inclined reader may want to sit down with pencil and paper (and eraser?) and try to construct one "by hand." It's not (so) easy. Indeed, it's not a priori obvious that there are such sequences for arbitrary values of k. In chapter 4, on universal cycles, we give very similar-sounding problems, where the sequences do exist for some values of k, but not for others.

We have thus arrived at a math problem: Given k, do there exist de Bruijn sequences of window length k? If so, how many are there, and how can we find them? In the rest of the chapter, we will answer these questions and then show how they are applied to our card trick.

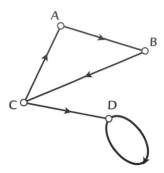

Figure 1. A simple graph

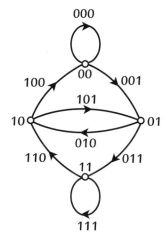

Figure 2. The de Bruijn graph on four vertices

One way of answering the question "Are there de Bruijn sequences for every k?" uses graph theory. A directed graph can be represented as a bunch of dots (the *vertices* of the graph) and a bunch of arrows between some of these vertices (the *edges* of the graph). For example, figure 1 shows a graph with four vertices (A, B, C, D) and five edges. The edge from D to itself is called a *loop*. When you first meet graph theory, it is hard to imagine that there is much more to say. In fact, it's a healthy field of research with several speciality journals, a dozen or so yearly conferences, and hundreds of professional graph theorists. The following example should show you why.

Our current problem is to see if there is always at least one de Bruijn sequence of window length k. Form a graph with vertices being the strings of zero/one symbols of length $k-1$ (so there are 2^{k-1} of them) and an edge going from vertex x to vertex y if there is a zero/one string of length k that has x at its left and y at its right. As with many ideas, this is best understood by example. For $k = 3$, there are four zero/one strings of length $k-1$: 00, 01, 10, 11. Figure 2 shows the de Bruijn graph on these vertices.

For example, there is an edge from 01 to 11 because there is zero/one string of length three, namely, 011, that starts 01 and ends 11. Each of the edges is labeled by a zero/one triple. A de Bruijn graph can be *contemplated* for any k. However, they get harder to draw.

An "Eulerian circuit" in a directed graph is a walk (following the arrows) that uses each edge exactly once and winds up where it started. For example (trace with your finger), starting at the bottom (vertex 11), visit 11 (again, by edge 111), then 10, 01, 10, 00, 00, 01, 11. If we write our steps down, separated by commas, the cycle is:

$$11, 10, 01, 10, 00, 00, 01, 11.$$

Since our walk follows the arrows, each vertex in the cycle has a common "center" with the following one. Collapsing our cycle by just indicating the new digit added gives a de Bruijn cycle:

$$1\ 1\ 1\ 0\ 1\ 0\ 0\ 0.$$

More generally, for any k, an Eulerian circuit in the de Bruijn graph gives a de Bruijn cycle with window length k.

This may not seem like much, trading a simple problem of zero/one strings for a difficult-to-visualize problem on an abstract graph. However, following on the great mathematician Leonhard Euler, we may easily see that a connected graph (i.e., you can get from any vertex to any other vertex by following arrows) has an Eulerian circuit if and only if each vertex has an equal number of edges leading in as leading out. For the de Bruijn graph, there are exactly two edges leading out of each vertex—a zero/one $(k-1)$-tuple can be finished off to a k-tuple in just two ways: add zero or add one. Similarly, there are exactly two ways of coming into a vertex. Furthermore, it is not hard to check (by changing one digit at a time) that we can go from any vertex to any other vertex along some path following the arrows. Since we have verified the conditions for Euler's theorem, we may use its conclusion: de Bruijn sequences exist for every k. The proof of the theorem even gives an algorithm of sorts for construction: Start at *any* vertex (say $k-1$ 0's), choose any available arrow leading out, erase this arrow, and continue. The proof shows you can cover each edge just once without getting stuck. What's more, the construction forces a cycle; the last step winds up at the original start. (Strictly speaking, you may end up with a number of smaller cycles that can then be stitched together to get one big [Eulerian] circuit.)

A bit more is coming, but let us return to the trick. By actually drawing the graph for $k=5$ (it has 16 vertices and is a bit of a mess), we find lots of de Bruijn sequences. One is:

$$00000100101100111110001101110101.$$

Let us use this to make a performable version of the trick. Get a deck of cards and remove all the aces through eights of all four suits (thirty-two cards in all). Arrange the cards in the following order:

8C,AC,2C,4C,AS,2D,5C,3S,6D,4S,AH,3D,7C,7S,7H,6H,
4H,8H,AD,3C,6C,5S,3H,7D,6S,5H,2H,5D,2S,4D,8S,8D.

This matches the zero/one string above—the colors start B,B,B,B,B,R,B,B, . . . The top card of the arranged stack is the eight of clubs (8C), the next card is the ace of clubs (AC), and the bottom card is the eight of diamonds (8D). The deck as arranged can be given any number of cuts. This does not change the cyclic pattern, only the starting point. To perform the trick, the performer must be able to "decode" the pattern prescribed by the five spectators and convert it into the names of the five cards. Here is a practical way of doing this. Table 1 lists the five cards next to each possible pattern.

Table 1. Possible card patterns

00000	8♣, A♣, 2♣, 4♣, A♠		01000	8♠, 8♦, 8♣, A♣, 2♣
00001	A♣, 2♣, 4♣, A♠, 2♦		01001	A♠, 2♦, 5♣, 3♠, 6♦
00010	2♣, 4♣, A♠, 2♦, 5♣		01010	2♠, 4♦, 8♠, 8♦, 8♣
00011	3♣, 6♣, 5♠, 3♥, 7♦		01011	3♠, 6♦, 4♠, A♥, 3♦
00100	4♣, A♠, 2♦, 5♣, 3♠		01100	4♠, A♥, 3♦, 7♣, 7♠
00101	5♣, 3♠, 6♦, 4♠, A♥		01101	5♠, 3♥, 7♦, 6♠, 5♥
00110	6♣, 5♠, 3♥, 7♦, 6♠		01110	6♠, 5♥, 2♥, 5♦, 2♠
00111	7♣, 7♠, 7♥, 6♥, 4♥		01111	7♠, 7♥, 6♥, 4♥, 8♥
10000	8♦, 8♠, A♣, 2♣, 4♣		11000	8♥, A♦, 3♣, 6♣, 5♠
10001	A♦, 3♣, 6♣, 5♠, 3♥		11001	A♥, 3♦, 7♣, 7♠, 7♥
10010	2♦, 5♣, 3♠, 6♦, 4♠		11010	2♥, 5♦, 2♠, 4♦, 8♠
10011	3♦, 7♣, 7♠, 7♥, 6♥		11011	3♥, 7♦, 6♠, 5♥, 2♥
10100	4♦, 8♠, 8♦, 8♣, A♣		11100	4♥, 8♥, A♦, 3♣, 6♣
10101	5♦, 2♠, 4♦, 8♠, 8♦		11101	5♥, 2♥, 5♦, 2♠, 4♦
10110	6♦, 4♠, A♥, 3♦, 7♣		11110	6♥, 4♥, 8♥ A♦, 3♣
10111	7♦, 6♠, 5♥, 2♥, 5♦		11111	7♥, 6♥, 4♥ 8♥ A♦

One way to use the table is to pencil it lightly on the top portion of a pad of paper (you can also photocopy it). Take the deck out of its case (don't forget the rubber bands). Have five cards selected as described. Pick up the pad and a felt-tipped pen (ostensibly to aid your visualization process). Make some scribbles on the pad as you patter about the spectators' powers of concentration. After admitting difficulty, ask the spectators holding red cards to stand up. Mentally translate this into a binary pattern, say 01001 (zero for black, one for red). Find this pattern on the list. You now know all five cards and can reveal them in a dramatic way, perhaps naming the red cards first and then the blacks.

It is important to give no clue that you are consulting a list. This can be helped by thought and practice. To begin, note that the first eight rows in the list contain patterns that start with 00, the next eight rows have patterns starting with 10, then 01, then 11. The upper half of each group of eight in the list contains patterns ending in 000, 001, 010, 011. The lower half contains patterns ending in 100, 101, 110, 111. When you see how the spectators stand up, your hands on the pad can locate the correct group of eight and the correct upper or lower half. This should be done without looking. Then, a glance down determines the exact pattern. Put a finger or thumb there. Now begin doodling on the pad, looking at the exact card names. You can write a few of the correct cards in large letters to end the revelation. There is no substitute for practice. Plan what you will say, keep talking, pretend you actually are a mind reader in a bit of trouble. We suggest fifty run-throughs as a minimum number required to perform this well.

One of our former students, now a professor himself, harnessed the computer to replace the list. He wrote a short program that takes five binary inputs and outputs the five chosen cards. He creates misdirection by asking for and inputting seemingly irrelevant data ("What country were you born in?", "Did you have orange juice for breakfast this morning?", etc.). He uses the computer for his "cheat sheet." The first person to make this into an iPhone app wins a free glass of orange juice from us!

At the end of this chapter, we give a way to completely eliminate any secret lists—the whole trick can be carried out mentally. This development is possible only because of some very elegant mathematics.

Where are we in our understanding of de Bruijn sequences? The mathematics of the de Bruijn graph shows that, in principle, we can always find a de Bruijn sequence. However, we don't have any concrete method in hand and, as will emerge, there are lots of different constructions that are useful for different applications.

One systematic approach is the "greedy algorithm." Begin by writing a sequence of k zeroes and then adding a one whenever you can (in other words, whenever you don't form a pattern that you have already seen). Thus, for $k = 4$, begin 0000 and cross this pattern off of the list. Adding ones (and crossing off the list each time) gives 00001111 as the first eight symbols. Adding another one would give a repeat, so a zero must be added instead. Continuing leads to the final sequence:

$$0\,0\,0\,0\,1\,1\,1\,1\,0\,1\,1\,0\,0\,1\,0\,0.$$

This rule works for $k = 4$, and M. A. Martin (1934) showed that the rule works for all k.[1] A practical person may only need to construct a sequence for a fixed value of k and wonder why a mathematician cares about all k. After all, *no* application will require a truly large

k (larger than one hundred, say) and, even for $k = 40$, for example, $2^{40} \approx 10^{12}$ is not difficult to try on today's computers. Why bother with a more careful analysis? While there is no explaining curiosity, we offer two questions that cry out for theory. In the card trick above, we are given k consecutive symbols and need to know where we are in the sequence. In applications in chapter 3, we have to do the opposite—given a position in a long de Bruijn sequence, what are the k following symbols?

Consider the first task. On the sequence generated from the greedy algorithm when $k = 4$, suppose we see 0110. Is the next symbol a one? It would be unless 1101 had been used up earlier. As it turns out, the next symbol is a zero. Thus, knowing the next symbol seems to require searching through a complete list of all the earlier occurring patterns. Of course, this is just a first thought on the matter. Maybe a more careful look at the greedy algorithm will reveal some useful structure. This is again a math problem. We can prove that, for large k, the storage list we require in order to look things up if the greedy algorithm is used must be exponentially long in k. Hal Fredricksen gives a clever variation of the greedy algorithm that requires only three times the window length k in storage. Below we discuss other constructions that make the "what is the next symbol" question easy.

Once one considers different methods of construction, it is natural to ask: "How many de Bruijn sequences with a fixed window length are there?" We consider two de Bruijn sequences to be the same if they differ only by a cyclic shift. Thus, for $k = 3$, it is easy to check that there are just two:

$$00011101 \quad \text{and} \quad 11100010.$$

For $k = 4$ there are 16. For $k = 5$ there are $2^{11} = 2,048$. De Bruijn got his name on the sequences by giving an amazing formula:

For any k, the number of de Bruijn sequences is exactly $2^{2^{k-1}-k}$.

We will leave further developments to the next section. For now, we have met de Bruijn sequences, shown that they exist, have given methods of construction, and we have been able to count them. As later chapters show, there are useful, natural variations where any of the questions of existence, construction, and counting are still open problems.

GOING FURTHER

The magic trick In Cycles is our version of a trick of Charles T. Jordan's. Writing in 1919, Jordan described Coluria, a trick with thirty-two cards repeatedly cut with a pattern of colors revealing a selected card.[2] We will tell Jordan's amazing story in chapter 10. He was a chicken farmer from Petaluma, California, who invented and sold card tricks. He made part of his living as a professional problem solver, entering newspaper contests of "impossible questions" in cities around the country. Despite these abilities, even Jordan couldn't quite get it right. For a deck with thirty-two cards, he asked for the colors of six consecutive cards! In the 1930s, the magical inventors William Larson and T. Page Wright marketed a trick called Suitability. Here, a deck of fifty-two cards is repeatedly cut and three cards are removed. The spectators announce the suits of their cards and the performer correctly names them. The number of possible answers is $4 \times 4 \times 4 - 64$, so there is enough information to distinguish fifty-two possibilities. The reader is invited to find a suitable arrangement. In chapter 4, on universal cycles, we give a general solution.

In the 1960s, Karl Fulves and, separately, P. Diaconis working with the chemist Ronald Wohl, derived dozens of tricks based on variations and extensions of Jordan's idea. Magicians have kept at it. They mistakenly call de Bruijn sequences "Gray codes." Indeed, there *are* combinatorial Gray codes, which are sequences of k-tuples, each differing from the last by changing one digit. For example, 000, 001, 011, 010, 110, 111, 101, 100 is a Gray code for $k = 3$. The difference is that Gray codes cannot be arranged into one sequence with consecutive k-tuples differing by a shift. The distinct blocks of k differ by one digit that may be changed in any place. Gray codes are extremely useful and interesting objects. They are used in analog to digital conversion, to calculate correlations, and in Samuel Beckett plays. But as far as we know, there has never been a single use in magic. (Now there is a magic problem!)

The reader wishing for a gentle introduction to graph theory and de Bruijn sequences can do no better than to consult Sherman K. Stein's marvelous works.[3] More advanced (but still friendly) treatments are given by Hal Fredricksen and Anthony Ralston.[4] A comprehensive treatment, with many topics not covered here, is in Donald Knuth's long-awaited *The Art of Computer Programming* 4A, part 1.[5] This covers de Bruijn sequences, Gray codes, and much, much more. An online encyclopedia on the subject is available from Frank Ruskey.[6]

AN ELEGANT SOLUTION TO THE MAGIC TRICK

In this section, we describe a way of getting rid of any lists in performing the trick. We also describe the full performance details. The solution involves working with the binary number system to count from zero to seven.

The usual way we write numbers is in base ten. Thus, 11 is $1 \times 10 + 1$, and 274 is $2 \times 100 + 7 \times 10 + 4 \times 1$. Binary numbers work with powers of two. Thus, 111 is $\mathbf{1} \times 4 + \mathbf{1} \times 2 + \mathbf{1} \times 1$ or 7, and 000 is $\mathbf{0} \times 4 + \mathbf{0} \times 2 + \mathbf{0} \times 1 = 0$. Similarly,

$$001 \text{ is } 1,$$
$$010 \text{ is } 2,$$
$$011 \text{ is } 3,$$
$$100 \text{ is } 4,$$
$$101 \text{ is } 5,$$
$$110 \text{ is } 6.$$

For instance, 110 is $\mathbf{1} \times 4 + \mathbf{1} \times 2 + \mathbf{0} \times 1 = 6$. Binary digits are *bits* (for BInary digiTs). The patterns of zeroes and ones we have been dealing with will be called "five-bit words." We will use the right-most three bits to denote one of the eight numbers zero to seven as in the list above. The left-most two bits will denote the suit. If the five bits are called *abcde*, we have

$$\underbrace{\text{suit}}_{a\,b} \quad \underbrace{\text{value}}_{c\,d\,e.}$$

The suit is coded according to the following rule.

00	club
01	spade
10	diamond
11	heart

Here, 0 in the left-most position denotes black, and 1 in the left-most position denotes red. We have used a standard bridge convention where hearts and spades are the major suits and diamonds and clubs are the minor suits, so that 1 in the second position denotes major and

0 denotes minor. Thus, 10 stands for the minor red suit diamonds. Most users will just memorize the four suit patterns.

This notation allows us to associate a card to each five-bit word, using the right-most three bits to denote value and the left-most two bits to denote suit. Thus, 00101 is the five of clubs. The stacked deck we introduce was derived from the sequence given on page 21 in just this way. Since there are no "zeroes" in a real deck of cards, we assign 000 to the numerical value eight (this is correct, modulo eight). Thus, the sequence begins 000001 . . . . The first five bits become 8C. The second five bits become AC, and so on.

Using this arrangement then automatically tells the performer the value and suit of the left-most spectator's card. The arrangement has been further designed so it is possible to tell the values of all five cards. To explain this, we need to introduce the operation of adding modulo two. This is a version of the well-known rule "even plus even is even, odd plus odd is even, while even plus odd is odd," familiar from adding numbers. If even is replaced by zero, and odd by one, we get the rules for adding modulo two:

$$0 + 0 = 0,$$
$$1 + 1 = 0,$$
$$0 + 1 = 1,$$
$$1 + 0 = 1.$$

These rules allow a simple description to get the next pattern of five from a starting pattern of five. The rule is this: If $abcde$ are five bits, the next bit is a plus c modulo two. Thus, 01001 is followed by 01001**0**. Translating to the language of cards, 01001 is AS. This is the left-most card. The card second from the left was determined from the five-bit word 10010.

first
⏜
0 1 0 0 1 0
⏝
second.

The second card from the left is therefore 2D.

From the original pattern, the name of the first spectator's card is known. Further, the next bit in the sequence can be computed and so the second spectator's card is known. The process can be continued to determine all of the bits. The rule is, to calculate the next bit in 010010, add the bits five back and three back from the end. For example, from **0**10**0**10 gives $1 + 0 = 1$, the boldface bits shown are five and three back. Adding gives $1 + 0 = 1$, so the next block of five is by 00101, which corresponds to 5C. The fourth and fifth cards are found to be 3S and 6D by following the same rules.

A sequence formed in the way described is called a *linear shift-register* sequence. Such sequences are used extensively in the mathematics of computer science. While we won't go into the theory of such sequences here, we point out a handy fact: Once you understand the rule, there is no real need to remember anything more.

Thus, suppose you are away from home and want to perform a card trick at a dinner party. The magically arranged list is not around. It's easy to create it: Just start with any non-zero five-tuple, say 00001. Using the rule, continue the sequence: 00001011 . . . , and then set the cards as the sequence demands. It is not even necessary to write out the sequence. Just get a deck of cards and remove the aces through eights. Start with any card, for example, the ace of clubs. Now use the rule: The next card is 00010, the two of clubs. Now it is easy to keep going, setting the cards in a few minutes. You will find the pattern runs through all non-zero five-tuples and so uses all cards *except* the eight of clubs. The pattern of thirty-one is just as before (see page 21), with the eight of clubs removed. This is the way we do the trick, using thirty-one cards.

We conclude this section with a few further notes on the practical performance of the trick without lists. To begin with, we have found that a few hours' practice enables rapid, sure calculation of the value of one card, given the next. Perhaps the best way to practice is simply to cut the deck, look at the top card, transform it into binary, use the rule to compute the next bit, and transform the last five bits into a card name. This can be continued.

Another approach uses the hands as a simple computing machine. In performing the trick, we observe the given pattern of reds and blacks among the five spectators. The idea is to generate the red/black pattern of the next four cards using the rule. To do this in an automatic way, use the first and second fingers of the left hand to represent the colors of the next two cards, and the first and second fingers of the right hand to represent the colors of the two following cards. This should be done without thinking. It requires assigning a definite order to fingers. Let a curled-in finger stand for a zero and a straight-ened finger stand for a one. Here is an example. The observed pattern is 01001. Using the rule, the next bit is zero. We curl in the second finger of the left hand. The next three bits are 1, 1, 0. These are suc-cessively represented by keeping the left first finger straight, the right first finger straight, and curling in the second finger of the right hand.

At this stage we have recorded 010010110. The five cards are thus AS, 2D, 5C, 3S, 6D. We have found this version straightforward to use in practice. There *is* some evident "computation" occurring in the mind of the performer, and this seems to enhance the effect of some-thing genuinely spooky going on. This is a nice example of a trick where the method is as amazing as the effect.

Magic performance offers opportunities not present in some of our other mathematical pursuits. For example, can we do Jordan's trick with a full deck of fifty-two cards where just five spectators are involved? Now, $2^5 = 32$ shows that the mathematical answer to the ques-tion is no. Any arrangement of the colors must have some repeated five-tuples. The magical answer is "Why not?" Find an arrangement of fifty-two cards with thirty-two distinct five-tuples and with twenty of these that repeat just once. Then, some of the time you know all the cards for sure and some of the time you know it is one of two fixed sets of five. A single further question will determine things. For example, "Spectator One, you have a red card. I think it is a heart." If yes, you are home. If no, you are home as well.

Throughout this book we go from magic to mathematics, and back. We have just posed a math problem arising from a magic trick: Is there a neat way to arrange a deck of fifty-two cards that does what is needed? We leave this to the interested reader.

Chapter 3

IS THIS STUFF ACTUALLY GOOD FOR ANYTHING?

The sequence 0000100110101111 has the property that successive groups of four 0000, 0001, 0010, 0100, . . . , go through each of the successive sixteen zero/one strings of length four once (going around the corner). Such a sequence is called a de Bruijn sequence of window length four. In chapter 2 we showed how longer versions with window length five are the basis of good card tricks. In this section we show how de Bruijn sequences and some variations are of use in robotic vision, in industrial cryptography, in putting together (and pulling apart) snippets of DNA, in philosophy, and in mathematics itself. The variations lead back to magic tricks, which in turn lead to new math problems, which we hope will lead to new applications.

ROBOTIC VISION

Picture an industrial robot going up and back in a long corridor. The robot changes direction as it senses activity. One design problem: the robot needs to know where it is. Instead of trying to keep track (let's see, 14 to the left, 77 to the right, 174 to the left, . . .) through difficult-to-measure turns of its wheels, Frank Sinden, a Bell Labs researcher, had the idea of labeling the path under the robot with a de Bruijn sequence. The robot can look down and report the zero/one string it sees (see figure 1).

Figure 1. A de Bruijn sequence for window length five

This practical solution to a real problem calls for very long de Bruijn sequences and a way of constructing them with a simple rule to convert back and forth between the zero/one patterns and positions in the corridor.

The real robotic vision problem is two-dimensional. Think of a robot tooling around the floor of a warehouse. To understand, consider the array:

1	1	0	1
0	0	0	1
1	0	0	0
1	0	1	1

If a 2 × 2 window is placed at the upper left-hand corner, it shows $\begin{smallmatrix}1&1\\0&0\end{smallmatrix}$. Sliding the window around anywhere, including going around the edges (or even corners), you will always find a different pattern. The 2 × 2 window ⊞ has four places, and each can be filled zero/one. This leads to sixteen different patterns. Each occurs once and only once to give a unique signature. Hence, the 4 × 4 array is a two-dimensional de Bruijn pattern.

A more recent application of these ideas has arisen in connection with so-called digital pens. By using special paper on which two-dimensional de Bruijn arrays have been invisibly imprinted, these electronic pens always know where they are on the page and can then be used in many amazing ways.[1]

Suppose you had to design a larger version. With a 3×3 window, there are $2^9 = 512$ configurations. With a 10×10 window, there are 2^{100} configurations, which is enormous. These are too big to do fooling around by hand. Some systematic scheme is needed. You might say, "I need a mathematician." Alas, if you pick a mathematician at random, you're likely to get "I don't do stuff like that" as an answer. Most mathematicians work with calculus-type "smooth" problems, not discrete things like cleverly arranged arrays of zeros and ones. Hopefully, your mathematician would say, "You need a combinatorialist." That's us (your two authors) and our friends.

Over the last fifty years, a number of people have gotten the "curious look" and had to think about these higher-dimensional de Bruijn arrays.[2] Also called de Bruijn tori, or perfect maps, they come in many flavors. Instead of a 2×2 window, we may need a 3×3 or 2×3 (or more generally, $u \times v$). Instead of zeroes and ones filling our window, red/white/blue (or, more generally, c colors) may be used. A 2×3 window may be filled with red/white/blue in $3^6 = 729$ ways. Is there a 27×27 array with each cell painted red, white, or blue such that each 2×3 window is distinct? As with de Bruijn sequences, there are a number of basic questions. Given an array size (say, $s \times t$), a window size (say, $u \times v$), and a number of colors (say, c), we may ask:

- Do de Bruijn arrays exist?
- If so, can an explicit construction be found?
- How many such de Bruijn arrays are there?
- Are there nice constructions (easy to construct, easy to find where you are)?

Almost all of these questions are open research problems as of this writing.

The two-dimensional de Bruijn arrays arose from practical problems. For us, it is natural to ask, "What's the trick?" How can one of those arrays be used as the basis of a magic trick? We still don't have a good answer but the story of the search opened up some new mathematics.

Let us begin with the problem of how to make a magic trick out of the two-dimensional de Bruijn array. In the spring of 1992, the authors taught a Mathematics and Magic course at Harvard University. During class discussion, the students came up with the idea of having a map with countries and pictures illustrating some of their properties (such as ski slopes or beaches). They thought of eliminating the 2 × 2 window using the following ruse: "This is a map of a secret land." The performer lays the map out on the table; it has a compass showing north-south-east-west. The performer turns his back to the audience and continues: "Would you collectively choose any one of the countries on the map and Rebecca, put your finger on the chosen country. I don't want to know the name of your country but just to get in the mood, tell me if the country to the north is hot or cold." They look and if it has sand and sun they say hot; if it has snow they say cold. The performer continues: "What about your neighbor to the south? Is it hot or cold? Now look at your neighbor to the east, hot or cold?" Finally, they tell the performer "hot or cold" for the western neighbor. This is four zero/one answers and amounts to using a window of the shape depicted in figure 2.

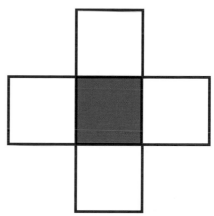

Figure 2. Cross window

They report a zero/one answer for each of the four squares shown. In principle, this should be enough to tell where the shaded cross is centered. We were excited! After all, this was the first and only trick ever with a two-dimensional de Bruijn array. We asked the students to rough out an actual layout of sixteen squares (with zeroes and ones now) that would work with the north-south-east-west window shown. At the next meeting, a glum group reported that they couldn't find a pattern that worked. After trying and trying, they finally found a proof that no such pattern was possible. Can the reader see why not? We pose a second problem to the reader: Is there a reasonable magic trick using a two-dimensional layout? It doesn't have to be what we showed in figure 2, just *some* genuinely two-dimensional geometry.

The story has a happy ending in another direction. We found the idea of using other window shapes new and fascinating. It is interesting even in one dimension. The de Bruijn sequence is defined via a sliding window of consecutive blocks. What about a "comb" in which the shaded blocks are opaque and the open blocks are transparent. For four open blocks, as shown in figure 3, one seeks a string of sixteen zeroes and ones with the property that if the comb shown is laid on top and slid along the string of sixteen (going around the corner), the visible zeroes and ones go through each of the sixteen visible patterns once and once only. We recommend the problem and remark that the theory of what can and cannot be done is in its infancy. We do not know useful conditions on the comb for the existence of solutions. Counting solutions seems far in the future. What little *is* known

Figure 3. A comb with spaces in positions 1, 2, 4, 8

Figure 4. A de Bruijn cycle for the (1,2,4,8)-comb in figure 3

is found in a paper by Cooper and Graham.[3] In figure 4 we show an example of a comb of size four and a corresponding de Bruijn cycle of length sixteen for this comb. A computation by Steve Butler showed that this de Bruijn cycle was unique up to cyclic permutations and interchanging zero and one. We have no idea how many there are for general combs.

The robotic vision story illustrates a practical use for de Bruijn sequences. This led to two-dimensional de Bruijn arrays. Our desire to invent a magic trick based on these led to the invention of combs or general shape windows. This leads to research problems that are very much unsolved. The whole gives a picture of the ebb and flow among applications, fun, and research, which we find typical and thrilling.

MAKING CODES

What can you do with math? The government knows. They employ thousands of mathematicians to make and break codes. Cryptography is big business in industry too. Our bank and credit card transactions are encrypted. Hackers work around the clock to break these codes. Cable channels and music files are often protected by codes so you can't enjoy them until you've paid.

There are many flavors and schemes of code. The requirements are different for a spy in the field trying to send a few lines of instruction and for the designer of a code to protect televised sports transmissions. Curiously, there is a common theme that appears in codes of both types: This is "add modulo two."

Roughly, a text message (in English, say) is converted into a string of zeroes and ones using standard rules shown in table 1.

Table 1. Conversion table

A	000001	I	001001	Q	010001	Y	011001	6	110110
B	000010	J	001010	R	010010	Z	011010	7	110111
C	000011	K	001011	S	010011	0	110000	8	111000
D	000100	L	001100	T	010100	1	110001	9	111001
E	000101	M	001101	U	010101	2	110010	.	101110
F	000110	N	001110	V	010110	3	110011	,	101100
G	000111	O	001111	W	010111	4	110100	*	101010
H	001000	P	010000	X	011000	5	110101	:	111010

Thus, HELP becomes 001000, 000101, 001100, 010000. This is usually strung together as 0010000001010011000010000 to make a message to be sent. Now, to encode this message, a corrupting string is added modulo two, one character at a time, with no carries. Thus, $0 + 0 = 0 = 1 + 1, 0 + 1 = 1 + 0 = 1$. For example, with the message above and the corrupting string 0110101011101001101101101, we get:

Message	0010000001010011000010000
Corrupting string	0110101011101001101101101
Coded message	0100101010111010100101

This final coded message is now transmitted (say, by Morse code or as a picture sent from an Internet café). It is easy for someone who knows the corrupting string to decode the message. He or she just writes the corrupting string under the coded message and adds modulo two. Since $x + x = 0$ modulo two for $x = 0$ or 1, the corrupting string cancels out:

Coded message	0100101010101010101010100101
Corrupting string	0110101011101001101101101
Message	0010000001000011000010000 = HELP

The point is, if you *do not* know the corrupting string, it is impossible to break the code since each symbol in the coded message is zero or one depending only on what the corresponding symbol in the corrupting string is.

Where do we get a good corrupting string? For routine applications, one standard string or perhaps a standard code book (or computer file) of strings can be used. The gold standard of security is the "one-time pad." This is a string generated by a truly random process such as radioactive noise from the blips of a Geiger counter. Both the sender and receiver have a copy of this corrupting string. This makes things completely secure. One problem with the one-time pad is that the sender must keep a copy of the corrupting string. If the sender is caught and searched, a long string of zeroes and ones is prima facie evidence. One of the most exciting, true spy books we have read is Leo Marks's *Between Silk and Cyanide: A Codemaker's War, 1941–1945.*[4] Marks was a top codemaker (and breaker) for the British during World War II. He was the chief designer of codes used by agents behind enemy lines. He had the agents' one-time pads printed on silk squares in invisible ink. When a line of corrupting string was used, it was cut off and burned. The cyanide in the title refers to cyanide pills, supplied to agents in case they were caught and facing torture.

What does all this have to do with card tricks? One answer is de Bruijn sequences. Instead of a one-time pad, one may generate a long sequence of zeroes and ones using the ideas explained in chapter 2. Say the sequence starts 0000000000001. Form the next symbol by adding modulo two to the last entry plus the entries two back and five back. The resulting sequence continues 11010011 . . . without cycling for another 2^{20} terms (which is more than a million). We started it at 0000000000001. The sender and receiver can agree to start at the sequence spelled out by the first few words of the lead article in the *New York Times* on some particular day. Essentially, this same scrambling scheme is used millions of times a day in cell phones as part of CDMA technology.

The ideas are used in many variations. Here is a small example. The window length three de Bruijn sequence 00011101 is formed from 001 by first appending the sum of the first and last symbol and then deleting the first symbol until this process leads back to 001. Thus, 001 → 011 → 111 → 110 → 101 → 010 → 100 → 001. Here, we have added the eighth digit zero at the beginning. To use this as a code, proceed as in table 2.

The message column shows the eight possible zero/one triples in the order produced by the de Bruijn sequence. (We have added 000 to the top of each column.) The corruption column shows the following string (in the order produced by the de Bruijn sequence). The coded message column shows the sum of the message and corruption entries modulo two. Note that every zero/one triple appears just once as a

Table 2. Encyrpting a three-bit message

Message	Corruption	Coded message
000	000	000
001	011	010
011	111	100
111	110	001
110	101	011
101	010	111
010	100	110
100	001	101

Figure 5. DES chip (courtesy of Wikipedia user Matt Crypto)

coded message. To decode the coded message, look it up and add the matching corruption term (modulo two) to recover the message.

These codes (usually on slightly longer words, e.g., length five) are often hardwired into computer chips into units called S-boxes. Several of these with different de Bruijn cycles in each are often hooked up together so that after a message passes through one S-box it is passed to a second—just which one depends on the message—then to a third, and so on, until a final scrambled string is output. With careful choices of de Bruijn sequences and box-to-box rules, these schemes make up a practical algorithm—the Data Encryption Standard (see figure 5). It was invented by researchers at IBM and the National Security Agency. For many years it was used millions of times a day for every kind of practical activity.

Here is a crypto story from the trenches. We were once employed by a Fortune 500 manufacturer to design a second-rate crypto scheme. The problem was export licenses. At the time, the U.S. government would not allow truly secure crypto systems (such as the Data

Encryption Standard) to be used in products sold overseas. The manufacturer wanted a crypto chip to protect special-event TV shows. They just wanted protection against hackers, not security against a sophisticated, high-level attack with all the computer power of the National Security Agency. We did our best, designing a complex scheme of sixteen S-boxes with linear de Bruijn sequences in each. To get an export license, we had to let a division of the government's spy service test our chip. We passed the test, which means our scheme wasn't really any good. After that, manufacturing began. It's a strange excitement seeing thousands of plug-in chips with a scheme you sketched out on a pad. The company made some sales and, after a while, the Data Encryption Standard was authorized for export and our system was retired.

Crypto is an important, big business now. Every U.S. corporation has its own experts, and crypto companies such as RSA (named after the computer scientists Ron Rivest, Adi Shamir, and Len Adleman) bill millions a year. It all goes back to zeroes and ones and the kind of schemes we have described above.

TO THE CORE OF OUR BEING

DNA is made up of sequences of four symbols (A, C, T, G) such as AACTCCAGTATGGC. . . . The patterns hidden in DNA strings are used to identify criminals, determine true parentage, understand diseases, and create cures. It is important stuff.

A sample of DNA is easy to get from a hair fragment, saliva, a semen stain, or a bone fragment. It is hard to read the associated strings. The story starts there. At the end, the mathematics behind our card trick underlies a promising technique for reading DNA.

Figure 6 shows a sequencing chip. This is an eight-by-eight array with a site for each of the $4 \times 4 \times 4 = 64$ patterns, from AAA to TTT.

A strand of DNA to be sequenced interacts with the array and every consecutive triple present in the strand is highlighted in the array. This highlighting is shown for the example AACTCCAGTATGGC (see figure 7). The problem is, given the highlights, what string occurred?

While there are many caveats (some spelled out below), we can also attack this problem by using the shared triples to form a de Bruijn graph. The vertices are the sixteen pairs from AA to TT. An edge is

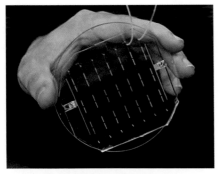

Figure 6. DNA sequencing chip (courtesy of Berkeley Lab, www.lbl.gov)

AAA	ACA	AGA	ATA	AAC	ACC	AGC	ATC
AAG	ACG	AGG	ATG	AAT	ACT	AGT	ATT
CAA	CCA	CGA	CTA	CAC	CCC	CGC	CTC
CAG	CCG	CGG	CTG	CAT	CCT	CGT	CTT
GAA	GCA	GGA	GTA	GAC	GCC	GGC	GTC
GAG	GCG	GGG	GTG	GAT	GCT	GGT	GTT
TAA	TCA	TGA	TTA	TAC	TCC	TGC	TTC
TAG	TCG	TGG	TTG	TAT	TCT	TGT	TTT

Figure 7. Array of triples from AACTCCAGTATGGC

drawn from one pair to another if there is a shaded triple that begins with the first pair and ends with the second. Thus, because AAC is shaded, the graph for these data has an edge from AA to AC. The full graph is shown in figure 8.

This looks like a mess. To reconstruct the sequence, we need a tiny variation of Euler's theorem; our previous version, discussed in chapter 2, gave a necessary and sufficient condition for a circuit, a path crossing every edge just once *and* starting and ending at the same place. For sequencing problems, the start and end don't have to be the same. The theorem says that a graph comes from a sequence starting at x and ending at y if and only if the in-degree of each vertex equals the out-degree of each vertex for all vertices except that $out(x) - in(x) = 1$ and $out(y) - in(y) = -1$ (and, as before, it is always possible to go from any vertex to any other vertex along *some* path). Looking at the graph, we see AA has in-degree one and out-degree two. So $out(AA) - in(AA) = 1$. Further, $out(GC) - in(GC) = 0 - 1 = -1$. All other vertices have out-degrees equal to in-degrees. From Euler's theorem, there is a sequence that starts AA, ends GC, and makes all needed transitions. Further, we can find the sequence by starting at the forced start AA and choosing any available arrow leading out. Each arrow is erased when used. If there are several arrows, use any available. The theorem says we never get stuck, and when finished we'll have a path starting at AA and ending at GC, with matching transition arrows.

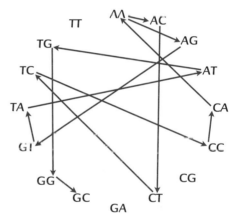

Figure 8. Full graph for AACATTACAATCACCGA

Strictly speaking, this process may also create cycles, but these can be easily merged into the final path.

The example shown is simple. For longer sequences, there may be several different reconstructions that match the available data. We have also suppressed problems of errors in highlighting and some other problems in order to explain the idea. This idea is the basis of a sophisticated algorithm, called *Euler*, which is used with much more complicated "shotgun" data. Here, a single strand of DNA is first replicated to make many exact copies. (This turns out to be easy to do with a chain reaction.) These copies are then snipped into pieces by a clever chemical process. For our example AACTCCAGTATGGC, this would yield pieces such as AACT, AA, AACTC, and so on, each repeated many times. Unfortunately, it also yields many copies of each of the "opposite" pieces where A ↔ T, C ↔ G, G ↔ C, T ↔ A are exchanged. For example, the opposite of AACT is TTGA. From this wealth of contiguous pieces and their opposites, the problem is to put together the original string. It is amazing this can be done at all. Doing it efficiently in a noisy environment is where the de Bruijn graph comes in. It would take us too far afield to give further details.[5]

We cannot leave this topic without talking about one of our inventions: an application of de Bruijn sequences to understanding DNA that started as a solution to a philosophy problem. Let us start with the DNA application. We have all heard things like "98 percent of our DNA is the same sequence as the DNA of a mouse." This leads to the notion of the distance between two DNA strings. If x = AACGCTT . . . and y = AATCTTG . . . are two DNA strings, a variety of distances $d(x, y)$ are used to measure how similar x and y are. These distances are used to align sequences, to measure similarity (is a man more similar to a chimp than to an ape?), and for finding the closest match to a new strand of DNA in a large database of sequences. Distances are often based on the minimum number of "moves" required to bring x to y. The allowed moves involve inserting and deleting characters and reversing a portion of the sequence. Thus,

$$\text{distance}(AAT, AAGT) = 1 \text{ by just inserting } G.$$

There are many more ideas involved in constructing useful distances. We will not give further details because they do not matter here. Thus, suppose we have picked a distance d and have two fixed strings x, y. We calculate $d(x, y) = 137$. The question is: "So what? Is this big or small?" One widely used answer involves "wiggling" x or y to get a comparison collection of related (but random) sequences. To fix ideas, consider the sequence

$$x = \text{AACATTACAATCACCGA}.$$

We construct a transition array for x, recording for each possible pair of letters how often it occurs as a block of two. For the sequence above, this gives:

	A	C	T	G
A	2	3	2	0
C	3	1	0	1
G	1	0	0	0
T	1	1	1	0

The entry in the row A, column A is 2 because AA appears twice in the sequence.

There are many different strings with this same transition array, for example, AATTACACCGAATCACA. The problem is this: Choose (repeatedly) random strings with a matching transition array. These random strings are used to calibrate the original distance. It turns out this is easy to do *if* you know about the de Bruijn graph. The idea is simple. Given the array, form the de Bruijn graph with four vertices A, C, T, G and an arrow from one vertex to another with weight equal to the entry of the array. For this example the de Bruijn graph is shown in figure 9.

To generate a random string starting at A, just pick one of the arrows leading out of A and follow it, deleting one from the weight on the edge. Write down A followed by whatever symbol you land on. Keep going in the same way, recording the vertices you pass through. One fine point: not all sequences with the same transition array start with the same symbol. There are several other applications of de Bruijn sequences to DNA that we don't have space to discuss.[6]

The example clearly connects DNA analysis to card tricks (via the de Bruijn graph). It is probably less clear what it has to do with philosophy. The connection is through de Finetti's theorem for Markov chains. Briefly, in trying to interpret classical probability statements such as "tossing a coin with unknown bias" from the viewpoint of

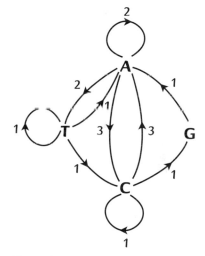

Figure 9. Labeled de Bruijn graph for AACATTACAATCACCGA

subjective probability, de Finetti invented the symmetry notion of *exchangeability*. This let him talk about symmetric events without having to mention unobservable things such as "coins with unknown bias." He then proved that his symmetrical allocations can be represented as a mixture of classical coins with a known bias. The weights in the mixture were called an "a priori distribution" by Bayes and Laplace.[7] In extending de Finetti's notions from exchangeability to partial exchangeability (required to give a subjective interpretation of Markov chains), we encountered the problem of having to understand a random string with a given transition array. Familiarity with de Bruijn sequences made this a friendly problem and thus a complete solution can be given.

Our understanding of de Finetti's theorem leads back to biology in another way. A group of chemists studying protein folding needed to discretize the physics of a huge dynamical system. They wanted their discretization to mirror the time reversibility of ordinary mechanics. The resulting "prior distribution on reversible Markov chains" was worked out in joint work with Silke Rolles.[8] Extensions to higher-order reversible chains are needed for validation. The de Bruijn graph was crucial here.

From card tricks, to DNA, to philosophy, and back to protein folding is a wide sweep for an idea. However, we want to show how this mathematical idea has many other uses. Our last section has quite a different flavor.

THIS DE BRUIJN STUFF IS COOL BUT CAN IT GET YOU A JOB?

Combinatorics sometimes seems to be about solving puzzles and asking riddles. It can be tricky but sometimes it doesn't seem like "real math." De Bruijn sequences are a perfect case in point. Can you really get paid to think about such stuff and, if you can, what kinds of things do you think about?

A snapshot of this kind of thinking occurred recently in an exotic location—a Banff resort in the Canadian Rockies. The Banff International Research Station is a mathematics institute that runs week-long conferences on focused topics.

Figure 10. The Banff International Research Station (photo by Scott Rowed, courtesy of The Banff International Research Station)

During the week of December 4–9, 2004, they ran "Generalizations of de Bruijn sequences and Gray codes." This brought together twenty-five or so researchers from all over the world. Brendan McKay came from Australia. He is a great combinatorialist who has achieved world-wide fame outside mathematics for his definitive debunking of the so-called Bible codes (claiming that there are hidden patterns in the Old Testament that can be used to predict the future). This particular week he was doing de Bruijn sequences. Eduardo Moreno came from Chile. He teaches at an institute that develops and applies de Bruijn sequences as part of its main mission. Robert Johnson, a fresh Ph.D., came from England.

The world's experts on our favorite subject were there—Hal Fredricksen, who teaches at the Naval Postgraduate School, did his thesis work on de Bruijn sequences and has developed them over a thirty-five-year period. Frank Ruskey from the University of Victoria has the world's best programs for generating all kinds of de Bruijn sequences and Gray codes on his Web site.[9] Carla Savage from North Carolina State University is an expert on complicated nonstandard constructions of amazing elegance. Also present were all levels of students—some were just starting and had taken a course that intrigued them. One hallmark of the subject: an outsider or newcomer can really make a contribution. Glenn Hurlbert from Arizona State University had recently finished a terrific Ph.D. thesis on generalized de Bruijn sequences.

What happens at such a conference? There were really friendly introductory and expository talks aimed at bringing newcomers up to speed and making sure we were all on the same page. There were announcements of new results, big and small. We did our card trick and talked about some variations on universal cycles (described in the following chapter). People talked about open research problems ("I'm stuck on this" or "I'm sure this is true but I just can't prove it"). Much time was spent in small groups where people go over special cases slowly and ask each other "silly questions" that might be embarrassing if asked in a large group.

One of the most spectacular new results was Robert Johnson's solution of the notorious "middle-layer" problem.[10] To explain, we have to augment our Eulerian circuits (paths in a graph that use each *edge* once and only once) with the notion of *Hamiltonian* cycles: paths in a graph that use each *vertex* once and only once. These Hamiltonian

Figure 11. Conference participants at the meeting in Banff (photo courtesy of The Banff International Research Station)

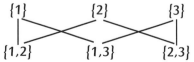

Figure 12. Graph of 1-sets and 2-sets of a 3-set

cycles are much more complicated to work with than Eulerian cycles. Indeed, if you can find a fast way to find out if a general graph has even one Hamiltonian cycle, it would really change the world. As an immediate consequence, thousands of other problems would be solved in one stroke. There is even a one million–dollar prize currently offered for a solution.[11]

Alas, we have often had to try to find Hamiltonian cycles for our card tricks (see the next chapter). Fortunately, these are for "nice neat graphs" instead of "general messy graphs," and we often succeed. Here is the problem that Johnson solved. Take an odd number n, say of the form $2r + 1$, such as 3, 5 or 1711584141. We will take $n = 3$ (so $r = 1$) for the moment. Consider the set of numbers $\{1, 2, \ldots, n\}$, so $\{1, 2, 3\}$ in our example. Form a graph with vertices being the subsets from $\{1, 2, \ldots, n\}$ of sizes r and $r + 1$. For $n = 3$, the subsets of size $r = 1$ are $\{1\}$, $\{2\}$, and $\{3\}$. The subsets of size $r + 1 = 2$ are $\{1, 2\}$, $\{1, 3\}$, and $\{2, 3\}$. Altogether, our graph looks like figure 12. In this figure, we have drawn an edge between a 2-element subset and a 1-element set if we can get the 1-element set by throwing out one number of the 2-element set.

The question/problem/conjecture is to decide if there is a Hamiltonian cycle in the graph of r-element and $(r + 1)$-element subsets of $\{1, 2, \ldots, n\}$ with an edge between an $(r + 1)$-element subset and an r-element subset if the r-element subset arises through the discarding of one number from the $(r + 1)$-element subset. In the $r = 1$ case above, we can trace the cycle

$$\{1\}\{1, 3\}\{3\}\{2, 3\}\{2\}\{1, 2\}\{1\}.$$

This uses each nonstarting vertex just once (of course, the starting vertex is the ending vertex). For $n = 5$ ($r = 2$), the graph is shown in figure 13. Can you find a Hamiltonian cycle in it? There is one (in fact, many). It has been checked by hand and computer that the conjecture is true for n up to twenty-nine. This middle-layer graph for $n = 29$ has 155117520 vertices and 1163381400 edges. This makes it difficult to do much hand-checking of possible Hamiltonian cycles! However, to this day, no one can prove that there is always a Hamiltonian cycle for middle-layer graphs with parameter n, for *all* n. Of course, Johnson can't prove that either (yet). What he did show is that the conjecture is *almost* true. He found a construction that works for all odd n and builds a path that meets all but a vanishingly small proportion of the vertices.

The problem is so simple to state that most of us at the conference had tried it. Moreover, the problem has been around for over fifty years, so many famous combinatorialists had tried, with very limited success. That the "kid" could come along and beat the world record by a mile is really exciting. Johnson's result introduces new ideas and techniques that will surely be of help in other graph-cycle problems. During his talk, we felt the wheels in our heads turn. As he spoke, people slowed him down as they formed their own mental picture of his new ideas. It must have been exciting for him too. He had an audience that really cared and wanted to follow. It was a wonderful hour.

The conference generated a lot of excitement. Brad Jackson and Glenn Hurlbert made progress on an old conjecture of ours (reported

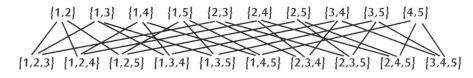

Figure 13. Graph of the 2-sets and 3-sets of a 5-set

in the next chapter). Many new conjectures were posed (and a few shot down in real time). Some conjectures posed were solved later. The main thing that happened is that we found a community; most of us are the only combinatorialist in our respective departments. To find others who think this small world of problems is beautiful and important made a deep impression on all of us. As a record, and follow-up, the progress we made and a list of open problems were collected together in an issue of the journal *Discrete Mathematics*.[12]

Can you make a living doing combinatorics? You bet you can—and have a lot of fun along the way, as well.

Chapter 4

UNIVERSAL CYCLES

We have used de Bruijn sequences for magic tricks and shown how they can be applied to make and break codes for spies and for analyzing DNA strings. The magic angle suggests wild new variations. Some of these lead to amazing new tricks. Some lead to math problems that will be challenges for the rest of this century. The following effect begins like the ones in earlier chapters but the information source is different.

ORDER MATTERS

Our start in this direction was the following magic trick, invented jointly with the chemist-magician Ronald Wohl. It has never been explained before. Here is how it looks. The performer tosses a deck of fifty-two cards, still in its case, out to an audience member. He or she tosses it to another, and so on, until everyone agrees it rests with a randomly chosen spectator. This person removes the deck and gives it a straight cut. The cards are passed to an adjacent spectator who gives it a cut. This is continued until a fifth spectator cuts the cards and takes off the current top card. The deck is passed back to the fourth spectator who takes the top card, and so forth, until the first spectator takes the top card. The performer points out that the deck is far out of his reach, and there isn't any way anyone can know which cards were taken. The patter continues, "Please look at your card and try to picture it in your mind. Help! You're doing a great job but I have a mess of images coming in. Would the person with the highest card please step forward (ace is high, two is low)?" One of the spectators steps forward. The performer looks pleased and scribbles a note on a pad. "Would the person with the next-highest card please step forward, and then the person with the next-highest card?" Now the questions stop: The performer names the three cards of the people standing and then names the cards held by the last two spectators.

Again, this is a trick that we have performed in the trenches. It plays well and really seems to fool people—especially other magicians. People just don't seem to think that the innocent patter and interaction provides enough information to give the names of all five cards.

How much information is there? In how many ways can the questions be answered? There are five spectators with *consecutive cards* from a randomly cut deck of cards. Any one of them can have the highest card; that's five possible answers to the first question. Any of the remaining spectators can have the next-highest card (four possible answers) and there are three possible answers for the third spectator. Together, this is $5 \times 4 \times 3 = 60$ possible answers. This is a bit more than enough to determine at which of the fifty-two positions the deck was cut.

Of course, the deck is arranged so that the performer knows the order of the cards and that there will be no ties. The arrangement assures that each consecutive group of five cards has a unique signature of "highest, next-highest, third-highest." Is such an arrangement possible? When we teach our mathematics and magic course we ask students to try to find an arrangement of an ordinary deck that does the job. While it is not easy, they can usually do it in an hour or less. We recommend it to the reader as a useful exercise. Recently, several students used a computer to solve the problem. We were startled when one of our students, Aaron Staple, told us, "It's easy. I just tried a few thousand at random and found several."

Let us try to explain (a) why this last finding surprised us and (b) how it leads into completely open waters. Consider first the odds of finding a zero/one de Bruijn sequence as discussed in chapters 2 and 3. Suppose we take a deck of thirty-two cards consisting of sixteen reds and sixteen blacks, shuffle it at random, and look to see if the resulting arrangement is a de Bruijn sequence. What is the chance we succeed? The number of possible arrangements of thirty-two cards is

$$32! = 32 \times 31 \times 30 \times \cdots \times 2 \times 1 = 263130836933693530167218012160000000.$$

From our previous remarks, we know there are $32 \times 2^{2^{5-1}-5} = 2^{2^4} = 2^{16}$ different de Bruijn sequences. We can label each of these with values in 16! × 16! ways. In all, then, the chance that a randomly shuffled deck of thirty-two cards forms a de Bruijn sequence is

$$\frac{16! \times 16! \times 2^{16}}{32!} \approx \frac{1}{9200}.$$

We thus expect to find one in about every 9200 trials; a million trials should produce more than a hundred different de Bruijn sequences.

This is a splendid example of how the computer has changed the way we think. In former days, trying something a million or so times was unthinkable. Now it's trivial and a skillful student can write a few lines of code and be done with it in half an hour or less. Doing the math (that is, discovering the structure of de Bruijn sequences) can take years.

On the other hand, the "try at random" strategy breaks down for longer sequences. For a window width of k and a deck of 2^k cards, the chances of a random shuffle landing on a de Bruijn sequence is

$$\frac{2^{2^{k-1}} (2^{k-1}!)^2}{2^k!}.$$

This is a difficult number to have a feel for. However, using Stirling's approximation for $n!$ it can be shown to be well-approximated by

$$\frac{\sqrt{\pi 2^{k-1}}}{2^{2^{k-1}}}.$$

This tends to zero super-exponentially fast.[1] As argued above, when $k = 5$ the chance of finding a de Bruijn sequence at random is about 1 in 9200. This is not so unlikely and, indeed, one of our students wrote a program that searched for random de Bruijn sequences for $k = 5$ and found hundreds of them. However, the odds decrease very rapidly as k increases, so that for $k = 6$ they are about 1 in 400,000,000 (still barely possible) and for $k = 7$ they are less than 1 in 10^{18}, which makes computation quite infeasible. Of course, if the algorithm used some intelligence rather than just a blind search, the odds could be increased substantially. However, with a random de Bruijn sequence, there is no simple coding as in our shift register method explained in the final section of chapter 2. Thus, there is still a healthy place for mathematics.

Let's get back to our "highest, next-highest, third-highest" card trick. Since a random computer search produced some examples, there must be plenty of them. As of the present writing we have *no* estimates, indeed no feeling at all, for how many (or how few) solutions exist. Our experience with zero/one de Bruijn sequences shows it is nice to have some theory.

Let us return to the effect that started this chapter and consider the neatest version of these ordered de Bruijn sequences. The problem is this: k distinct things can be arranged in $k! = k(k-1)(k-2)\ldots2\times1$ ways. Thus, three distinct things can be arranged in $3! = 3\times2\times1 = 6$ ways. Here they are:

$$123\quad132\quad213\quad231\quad312\quad321.$$

Is there an arrangement of the numbers 1, 2, 3, 4, 5, 6 so that, if consecutive groups of three are considered, the ordering of each group (high, medium, low) is distinct (we include going around the corner)? The following arrangement works:

$$1\ 4\ 6\ 2\ 5\ 3.$$

The first group of three, 146, is in the order low, medium, high (abbreviated LMH). The next group, 462, gives MHL. Then HLM, then LHM, then, going around the corner, HML, then MLH. Can the reader find an arrangement of the numbers 1, 2, 3, . . . , 24, such that each consecutive group of four gives a distinct order?

After experimentally finding sequences that work for groups of three, four, and five (this last involved a deck of $5! = 5\times4\times3\times2\times1 = 120$ cards), we became convinced that such "ordered" sequences always exist. The problem then became a math problem. We enlisted the mathematician Fan Chung to our cause and eventually showed that indeed they did always exist. Our proof was difficult (we needed to solve a Hamiltonian cycle problem). Further, our proof wasn't really constructive; we can't give a simple recipe, we just know there is always at least one way to do it. This solves the first problem on a list of problems. It leaves open the problem of finding a usable construction, an approximate or exact count of how many solutions there are, and the problem of finding an invertible construction (meaning two things— first, given a pattern of k, where does it lie within the long sequence and second, given a place in the long sequence, what pattern starts there?). Permutations of k things are such well-studied mathematical objects that there must be elegant answers to these questions. The applied successes of zero/one de Bruijn sequences offer hope that these new ordered sequences will find uses far beyond magic.

We continue our "order matters" study by describing three variations: real decks, repeated values, and products. Each offers similar challenges (and opportunities).

In a regular deck of cards, values are repeated. We have four aces, four twos, and so on. Any solution to the card trick that began this chapter has to deal with this problem. For example, consider a deck of twenty-four cards containing five cards labeled one, five cards labeled two, five cards labeled three, five cards labeled four, and only four cards labeled five. If they are arranged in the following order,

$$123412534153214532413254,$$

then every consecutive group of four has a distinct order. You can show that if the deck has six each of values one, two, three, and four, no such arrangement is possible.

For a given group size k, what is the smallest number of distinct values that can be assembled into a deck of $k!$ cards so that the relative order of each group of k is distinct? An old conjecture of ours was that this smallest number is $k + 1$, and we had offered a prize of one hundred dollars to anyone who could prove (or disprove) it. We are happy to report that this has now been proved in a very nice paper of Robert Johnson.[2] The next challenge is to design a nice magic trick based on this result.

The above deals with repeated values by sidestepping them, making sure that each consecutive group has all distinct values. Another approach is to *exploit* repeated values. Picture our card trick again. The deck of cards has been cut, people have taken consecutive cards, and the performer asks, "Will everyone having the highest value stand up?" Thus, one or more people may stand. The same question is repeated for the next value and so on. This gets us to the subject of permutations with ties. Consider three symbols, 1, 2, and 3. While there are only six arrangements of distinct values, if ties are permitted this increases to thirteen arrangements:

111, 112, 121, 211, 122, 212, 221, 123, 132, 213, 231, 312, 321.

Here, 111 is the same as 222 or 333 since all values are tied. If everybody with the highest card stands up, all three will stand and nothing distinguishes them further.

One may ask, can we arrange thirteen cards so that consecutive groups of three run through all possible permutations with ties permitted? The answer is yes:

$$5\ 5\ 5\ 4\ 5\ 1\ 3\ 2\ 2\ 3\ 2\ 1\ 4.$$

How many permutations of k things with ties are there? Call this number $G(k)$. We have $G(2) = 3$, $G(3) = 13$, $G(4) = 75$, $G(5) = 541$. The numbers grow rapidly. Can we arrange the usual deck of fifty-two cards so that each consecutive group of four runs over distinct permutations (ties permitted)? We don't know for sure (but we suspect the answer is yes). All of the open questions for ordered sequences are also open for permutations when ties are permitted. We would also like help picturing these things. We have tried to draw pictures of formally dressed permutations wearing ties or standing next to people from Thailand with limited success.

Our last development of "order matters" returns to practical card tricks. It *also* opens up many mathematical challenges. We call the topic Products but will explain it as a practical card trick. A deck is tossed out, cut many times, and three people take consecutive cards. The patter goes as follows: "You're doing a wonderful job of concentrating on your cards but there are too many images coming in. Let's see, would you rearrange yourselves so the person with the highest card is on the left and the person with the lowest card is on the right? Thanks. That's better. Please concentrate. Hmmm. Still a mess. I see red much more clearly than black. Would everybody with a red card please step forward?" Now the performer sees clearly and names the three cards.

There are six possible rearrangements of the three participants, and the red/black pattern can be rearranged in eight ways. In theory, this is $6 \times 8 = 48$ answers. Suppose the four aces are removed from the deck. Can we arrange the remaining forty-eight cards so that every group of three gives a distinct answer? Similar questions arise for combining any of the patterns above (or below). The theory of products is in its infancy but we can prove that products of any kind of de Bruijn sequence (e.g., zero/one or red/white/blue) can always be found.[4] A solution of the forty-eight-card-trick problem, courtesy of two of our freshman students, appears in table 1.

A MIND-READING EFFECT

The easiest (and perhaps best) trick to perform based on the product idea was developed by Ronald Wohl. Here is a brief description. An

Table 1. Sequence for "cut repeatedly, order yourself, and all red cards step forward" (thanks to Matt Duhan and Rebecca Rapoport)

AH	000	123	8D	000	231	2D	000	132
6H	001	231	JD	001	312	QD	001	321
QH	010	312	3H	010	123	8H	010	213
3C	101	123	9S	101	231	3S	101	132
7D	011	231	TD	011	312	JH	011	321
8S	111	312	AS	111	132	4S	111	213
4C	110	123	7C	110	213	2S	110	132
6S	100	231	5S	100	132	6C	100	321
9D	000	312	TH	000	321	5D	000	213
2H	001	123	9H	001	213	4H	001	132
5H	010	231	AD	010	132	7H	010	321
JS	101	312	TC	101	321	5C	101	213
4D	011	123	6D	011	213	3D	011	132
TS	111	231	AC	111	123	QC	111	321
QS	110	312	8C	110	231	9C	110	321
2C	100	123	JC	100	213	7S	100	312

Note: The first column shows the order of the deck. The middle and last columns show the color and order pattern if the corresponding card is cut to the top.

ordinary deck of fifty-two cards is handed out as above. Three spectators are enlisted, the deck is cut by each, and three consecutive cards are removed, one per spectator. The performer says to the first spectator, "As a warm-up, I want to try an easy test with you. Tell me the value of your card, but not the suit. It's only one out of four possibilities, but we hardly know each other." Say the spectator answers, "I have an eight." The performer answers, "Hold on. I want to work with all three of you at once, like a chess expert." To the second spectator, "Tell me the suit of your card only, not the value. That's a one-in-thirteen problem." Say the spectator answers, "I have a heart." To the third spectator, "I'll save the hardest for last. I don't want you to tell me anything, just concentrate on your card."

Continuing, the performer says, "Spectator one, you chose an eight. Try not to react, it could have been any of the four eights—clubs, hearts, spades, or diamonds; you shifted a little when I said 'clubs.' I think you have the eight of clubs." The spectator holds the card up to show that the performer is correct. To the second spectator, "I know you have a heart (if you'll pardon the expression!), but which one? High, medium, low? Face card or spot? Even or odd? Aha, I see; it is the king of hearts." The spectator holds up the card to show the guess is correct. Finally, after similar banter, the performer correctly names the third spectator's card without asking a single further question—"It is the three of spades."

HOW IT WORKS

The working of this trick is diabolically simple. First of all, the first spectator can give any of thirteen answers, the second spectator can give any of four answers, and 13 × 4 = 52, so there is enough information to know where the deck has been cut. We work with a deck arranged in the classical "Eight Kings" order, which follows the ditty, "Eight kings threatened to save ninety-five queens for one sick knave." That is, 8, K, 3, 10, 2, 7, 9, 5, Q, 4, 1, 6, J (eight [8] kings [K] threatened [3, 10] to [2]. . . .). The suits are arranged in "CHaSeD" order: clubs, hearts, spades, and diamonds, repeated cyclically. This determines an easy-to-remember order for the fifty-two-card deck, beginning with the eight of clubs on top, the jack of clubs in position thirteen, the eight of hearts at fourteen, and so on, through to the jack of diamonds at the bottom of the deck.

This time, as opposed to our previous versions, the working is easy. The first spectator tells you a value (here, an eight) and the second spectator tells you a suit (here, hearts). Since clubs always come before hearts (working backwards in the CHaSeD order) you know the first person has the eight of clubs. Since eights are always followed by kings, the second spectator must have the king of hearts and the third spectator must have the three of spades.

Any repeated cyclic order can be used. Another classic is known as the "Si Stebbins" order. Here, each value follows the last by adding three modulo thirteen. Thus, the values start 1, 4, 7, 10, K, 3, 6, 9, Q, 2, 5, 8, J, and repeat cyclically. You can use any memorized order, cyclic or not. This trick was invented by Ronald Wohl in the 1960s when we were working on variations of Jordan's Coluria. Since cyclic stacks were around well before 1600 and, in the end, no math is needed, it could have been invented earlier. This shows the power of mathematical thinking. Bravo, Ronald!

UNIVERSAL CYCLES AGAIN

We can finally explain the title of this chapter. Given *any* list of natural combinatorial objects, for example, zero/one strings or permutations, we can try to find a long (cyclic) sequence and a group length k so that each consecutive group in the sequence codes a unique object on our list. The long sequence is called a *universal cycle* (or *U-cycle*, for short). The field of combinatorics is filled with interesting objects. Finding U-cycles for any of them makes a sea of problems. Of course, the tricks involved may start out as pretty contrived efforts. With work, one can sometimes correct this and find a performable version.

Following Jordan's Coluria trick (described in chapter 2), the attorney William Larsen and the dilettante T. Page Wright introduced Suitability, in which a deck of fifty-two cards is cut, three people take consecutive cards, and each announces his or her suit (e.g., clubs, hearts, spades, or diamonds). Thus, there are $4 \times 4 \times 4 = 64$ possible answers, so it is possible for the performer to know the names of all three cards. This calls for a sequence of C, H, S, D of length fifty-two such that three consecutive cards all have distinct patterns. The computer expert Alex Elmsley marketed a version called Animal-Vegetable-Mineral. Here, a deck of twenty-seven cards with pictures of various objects is freely cut, three people take consecutive cards, and the game of twenty questions is played. The performer asks each spectator in turn if his or her card's object is "animal, vegetable, or mineral." This information reveals all three. We will hear more about Elmsley in the final chapter.

These variations are quite vanilla-flavored by the standards of this chapter. Let us take the reader through a mystery that really pleases us. The objects involved are the Bell numbers $B(n)$. This is the number of ways that n distinct things can be arranged into groups. For example, $B(3) = 5$ because three things, call them A, B, C, can be grouped as:

{A, B, C}	{A}{B, C}	{B}{A, C}	{C}{A, B}	{A}{B}{C}
All	A	B	C	All
together	apart	apart	apart	separate

Here, order within a group or among groups doesn't matter, so {A}{B, C} is the same as {A}{C, B} or {B, C}{A}. Just the groupings matter. The reader will understand better if he/she verifies B(4) = 15.

For years, we have been intrigued by the numerical coincidence

$$B(5) = 52.$$

We looked at each other and asked: "What's the trick?" After brooding about this (on and off, to be sure) for years, we finally saw a trick. You may want to think of your own before reading what we came up with. If you like puzzles but don't feel like waiting, don't worry, there are plenty of similar problems coming for which we don't know the answer.

THE EFFECT

As befits a good magic trick, we first thought of the effect and then worried about the method. Here is the effect. A deck of cards is handed out, cut many times, and then five people take consecutive cards. The performer asks them all to concentrate and complains that they are doing so well that a jumble of images is coming in. The patter continues, "I think we can do this together but I need your help. To reinforce your thoughts, I'd like you to group together. Would all the hearts stand together, all the clubs, and so on; people with the same suits stand together. Don't tell me your suits or anything else. Just concentrate." With no further questions, the performer names all five cards. The idea is, the spectators form into groups—this gives a partition of five. The cards should be arranged so that each consecutive group of five gives a different partition. For example, if the five cards are:

spectator 1	spectator 2	spectator 3	spectator 4	spectator 5
8C	4D	JD	AH	10C

the spectators would form the grouping {1, 5}{2, 3}{4}. Having the idea is one thing. With keenest interest we began to study its feasibility. What is needed is an arrangement of the symbols C, H, S, D in a row of length fifty-two so that each partition occurs just once. One problem: There is no way to have five separate groups of size one if only four suits are used. There are two easy ways around this: Work with fifty-one cards so one of the groupings isn't needed, or exchange one card for a joker, which serves as a fifth suit. Either way works. For the second option, using Hamiltonian cycles and a lot of hard work done jointly with Fan Chung, we found:

DDDDDCHHHCCDDCCCHCHCSHHSDSSDSSHSDDCHSSCHSHDHSCHSJCDC.

We can use this cycle with an ordinary deck with one of the spades replaced by a joker (= **J**). We will not give further details about practical performance; the interested reader will have to go forward alone. However, having worked through it ourselves, we can guarantee a superb, exclusive, performable trick that will astound the audience and please you. We know very little about constructions or the number of such Bell cycles. We are happy to have found one and look forward to seeing someone perform our trick.

Here is a last example where we know both more and less. It is a basic combinatorial object where some math has been developed, but we just cannot frame a reasonable trick. The combinatorial objects are subsets of $\{1, 2, \ldots, n\}$ of a given fixed size k (these are known as k-subsets of an n-set). For example, the 2-subsets of $\{1, 2, 3, 4, 5\}$ are:

$$\{1, 2\}, \{1, 3\}, \{1, 4\}, \{1, 5\}, \{2, 3\}, \{2, 4\}, \{2, 5\}, \{3, 4\}, \{3, 5\}, \{4, 5\}.$$

Thus, there are ten of them. Here $\{1, 2\}$ is the same as $\{2, 1\}$. Order doesn't matter. Can the reader find a sequence of length ten using the numbers 1, 2, 3, 4, 5 (each twice) so that each consecutive pair appears just once when order is ignored?

These k-subsets of an n-set are often called "combinations," the number of ways of choosing a committee of size k from a group of n people. They are a mainstay of classical combinatorics. A poker hand of size five is a 5-subset of fifty-two, and computing the odds for all kinds of games requires a thorough familiarity with such things.

The number of k-subsets of an n-set is used often enough that a special notation is common: $\binom{n}{k}$, read "n choose k." Thus, $\binom{5}{2} = 10$ from our list above. It is easy to see that $\binom{n}{k} = \frac{n!}{k!(n-k)!}$ with $n! = n(n-1)$ $(n-2) \cdots 1$. Thus, $\binom{5}{2} = \frac{5!}{2!3!} = \frac{5 \times 4 \times 3 \times 2 \times 1}{(3 \times 2 \times 1)(2 \times 1)} = 10$. Simple as all of this sounds, there are many mysteries hidden here: If you knew all about the primes dividing $\binom{2n}{n}$, then you would know a lot more than mathematicians currently know. For example, if we look at the first ten values of $\binom{2n}{n}$, we get the results shown in table 2. Notice, for example, that all the values are even numbers, that is, divisible by 2. In fact, this is true for *all* values of n. Can the reader see why this is true? However, there are values that are *not* divisible by 3, such as 20 and 70. Also, there are

Table 2. First ten values of $\binom{2n}{n}$

n	$\binom{2n}{n}$
1	2
2	6
3	20
4	70
5	252
6	924
7	3432
8	12870
9	48620
10	184756

values that are not divisible by 5 (such as 6 and 252) and some not divisible by 7 (such as 6, 20, and 3432). It is much harder to find values of $\binom{2n}{n}$ that are not divisible by *any* of the primes 3, 5, or 7. The first two are 2 and $\binom{20}{10} = 184756 = 2^2 \times 11 \times 13 \times 17 \times 19$. Can the reader find the next one? A well-known unsolved problem (for which the authors offer one thousand dollars to the first solver) is whether there is an unlimited number of such n.[3]

Back to our project. The problem is this: Given n and k, find a sequence of length $\binom{n}{k}$ consisting of symbols from $\{1, 2, 3, \ldots, n\}$ such that each consecutive group of k contains a different k-subset. Sometimes it can be done. For example, if $n = 8$ and $k = 3$, then $\binom{8}{3} = 56$, and we found:

82456145712361246783671345834681258135672568234723578147.

Our first surprise: You *cannot* always do it. It is *impossible* unless k exactly divides $\binom{n-1}{k-1}$. For example, if $n = 4$ and $k = 2$, then $\binom{n-1}{k-1} = \binom{3}{1} = 3$ so the impossibility result says it cannot be done. Let's see why. There are six 2-element subsets of $\{1, 2, 3, 4\}$. These are

$$\{1, 2\}, \{1, 3\}, \{1, 4\}, \{2, 3\}, \{2, 4\}, \{3, 4\}.$$

Suppose we had a valid sequence of length six, say, *abcdef*. Each of *a, b, c, d, e, f* stands for one or another of the symbols $\{1, 2, 3, 4\}$. Symbol 1 occurs someplace. The adjacent positions to this one have to be two other numbers giving two of the pairs containing 1. Now, 1 must appear twice and the new 1 gives two more pairs containing 1. This forces four pairs containing 1 in our proposed sequence. But there can be only three such pairs, so one of the four must repeat, violating the rules. Thus, no such *abcdef* exists.

The argument above shows that k dividing $\binom{n-1}{k-1}$ is a necessary condition. If k doesn't divide $\binom{n-1}{k-1}$ then there can be no solutions. The argument offers no help if k *does* divide $\binom{n-1}{k-1}$. It is not hard to see that these cycles exist when $k = 2$ and n is odd. It is easier to handle $k = n - 1$, since $n - 1$ always divides $\binom{n-1}{n-2} = n - 1$ and such sequences should (and do) always exist. The first interesting case we found beyond this was when $k = 3$. We couldn't figure it out. The problem became part of an academic talk we went around giving on mathematics and magic tricks. We gave it at Reed College and a visitor, Brad Jackson (now a professor at San Jose State University), became intrigued. He proved that the cycles exist for all n when $k = 3$ (and 3 divides $\binom{n-1}{2}$). His argument also worked for $k = 4$. The matter rested there for a while until our Ph.D. student Glenn Hurlbert showed that the required cycles exist for $k = 6$. This left $k = 5$ (and all larger k's) open. Recently, Hurlbert and Jackson combined forces and settled $k = 5$. The arguments are clever, long, and hard. They involve a lot of computer work.

With this effort in hand, what's the trick? We don't have a good answer but offer two bad tricks in the hope that some reader will get mad at us (or the problem) and invent a better trick.

To fix our ideas, consider the case $k = 3$ and $n = 8$ with $\binom{8}{3} = 56$. Above, we gave a sequence of length fifty-six made from the symbols $\{1, 2, 3, 4, 5, 6, 7, 8\}$ containing each 3-element subset just once. Such a sequence repeats each symbol seven times (to make up a deck of $7 \times 8 = 56$ cards). With any such arrangement, if a deck of cards is so set up, it can be cut any number of times, and the top k cards can be removed and shuffled randomly. Their unarranged values now determine the position of every card in the deck. One possible trick (remember, we're just brainstorming): A deck of fifty-six cards is made up containing eight strange symbols, with one symbol per card:

$$\sharp \ \blacksquare \ \Diamond \ \lll \ \text{⋔} \ \rightsquigarrow \ \text{⋔} \ \bowtie$$

Each symbol appears on seven cards. The deck is displayed and explained to the spectator. The performer brings out a personal computer that will take over from here. The computer displays the following instructions:

> Please put the deck on the table and cut it three times with your left hand. Remove the top three cards and shuffle them thoroughly. Lay them face-up in a row on the table [let's say they are $\blacksquare$ ⋔ $\Diamond$]. Key these in by touching the displayed symbols.

The computer now asks the spectator to choose one of the three as a "key" (let's say she chooses $\Diamond$). It then instructs the spectator to pick up the deck and follow a series of instructions:

> Deal into two piles alternately, put the top card of the left-hand pile aside. Pick up the right-hand pile and deal into two piles. Set aside the last card dealt.

The instructions continue in a seemingly "mad" pattern. At the end, six cards have been set aside face-down. The computer recalls the spectator's choice of ♢. When the six cards are turned up, they are the remaining ♢s.

This would take some work: For every cut position and choice of key, a dealing sequence would have to be entered into the computer's memory. It would make a nice class project (in which each student would be assigned a handful of cases). It's not the worst, but there must be a better trick.

A different thought: We can indicate a subset of three-out-of-eight whereby the red cards are in a row of eight. Does a sequence of fifty-six zero/one symbols exist such that each consecutive group of eight contains just three ones and their left-to-right positions run through each possible 3-subset just once? Well, a moment's thought shows that this is *not* possible since the only way to have three ones in *every* possible position of the sliding window is for the sequence to be periodic, with every symbol the same as the one eight positions earlier. However, perhaps we can arrange to have all the 3-subsets and 4-subsets appearing just once in a suitably arranged deck of $\binom{8}{3} + \binom{8}{4} = 56 + 70 = 126$? We have no idea!

Chapter 5

FROM THE GILBREATH PRINCIPLE TO THE MANDELBROT SET

One of the great new discoveries of modern card magic is called the Gilbreath Principle. It is a new invariant that lets the spectator shuffle a normal deck of cards and still concludes in a grand display of structure.

One of the great new discoveries of modern mathematics is called the Mandelbrot set. It's a new invariant that takes a "shuffle" of the plane and still concludes in a grand display of structure.

The above is wordplay; the connections between the invariants of a random riffle shuffle and the universal structure in the Mandelbrot set lie far below the surface. We'll only get there at the end. This chapter gives some very good card tricks and explains them using our new "ultimate" Gilbreath Principle. Later in this chapter, the Mandelbrot set is introduced. This involves pretty pictures and some even more dazzling universal properties that say that the pretty pictures are hidden in virtually every dynamical system. We'll bet you can't yet see any connection between the two parts of our story.

Right now, let's begin with our tricks.

THE GILBREATH PRINCIPLE

To try the Gilbreath Principle, go and get a normal deck of cards. Turn them face-up and arrange them so that the colors alternate red, black, red, black, and so on, from top to bottom. The suits and values of the cards don't matter, just the colors. With this preparation, you're ready

Figure 1. A riffle shuffle

to fool yourself. Give the deck a complete cut, any place you like. Hold the deck face-down as if your are about to deal cards in a card game. Deal about half the deck face-down into a pile on the table. The actual number dealt doesn't matter; it's a free choice. You now have two piles, one on the table, one in your hands. Riffle shuffle these two piles together. Most people know how to shuffle (see figure 1). Again, the shuffle doesn't have to be carefully done. Just shuffle the cards as you normally do, and push the packets together.

Here comes the finale: Pick the deck up into dealing position, and deal off the top two cards. They will definitely be one red/one black. Of course, this isn't so surprising since it happens half the time in a well-shuffled deck. Deal off the next two cards. Again, one red/one black. Keep going. You'll find each consecutive pair alternates in color. In a well-shuffled deck, one might naively expect that this would happen about $\frac{1}{\overbrace{1/2 \times 1/2 \times \cdots \times 1/2}^{26\,\text{terms}}}$ of the time (which is less than two chances in a hundred million). In fact, the odds are actually somewhat better than that, namely, about one chance in seven million. We will explain how we arrive at this number at the end of this chapter.

Before proceeding, you might want to figure out how it works. It's pretty easy to see that no matter how the cards are cut, dealt, and shuffled, it's a sure thing that the top two cards are one of each color. When we try this out on our students, it's quite rare for anyone to be able to see why the next pair is red/black. We don't recall a single student providing a full, clear argument for the whole story.

What is described above is called Gilbreath's First Principle. It was discovered by Californian Norman Gilbreath, a mathematician and lifelong magician, in the early 1950s. We'll have more to say about

Gilbreath at the end of this chapter. The second and ultimate principles are coming.

The red/black trick can be performed as just described. Thus, set up a deck of cards alternating red/black and put them in the card case. Find a spectator and proceed as directed above. Take the cards out of the case. Ask the spectator to give the deck a few straight cuts, deal off any number of cards into a pile on the table, and then riffle shuffle the pile on the table with the pile still in hand. As this happens, you might appear to be carefully studying the spectator (you might even fake making a few notes on a pad). You can promise that "This is an ordinary deck, not prepared in any way." Take the deck of cards and put them under the table. Say that you're going to try to separate the cards by sense of touch: "I promise I won't look at the cards in any way. You know, red ink and black ink are made up of quite different stuff. It used to be that red ink had nitroglycerine in it. Guys in prison used to scrape if off the cards. Anyway, I'll try to feel the difference between red and black and pair them up."

All you do is take the cards off the top in pairs. Pretend to feel carefully and perhaps occasionally say, "I'm not sure about these," and so on. If, as you display the pairs, you order them so each has a red on top followed by a black (you'll find they appear to come out in random order), assemble the pairs in order and you're ready to repeat the procedure instantly.

To be honest, the trick as just described is only "okay." It's a bit too close to the surface for our tastes. Over the years, magicians have introduced many extensions and variations to build it up into something terrific. As an example, we'll now describe a fairly elaborate presentation developed by the magician and insurance executive Paul Curry. The following unpublished creation has many lessons.

The performer gets a spectator to stand up and asks two quite personal questions: "Are you good at telling if someone else is lying? If you had to, do you think you could lie so that we couldn't tell?" It's a curious asymmetry of human nature that a large number of people answer yes to both questions (we owe this observation to Amos Tversky).

The props for this effect are a deck of cards and a personal computer (used as a score keeper). The performer asks the spectator to cut and shuffle the cards. Then two piles of ten or so cards are dealt off. The performer takes one of them and calls out the colors for each card, red or black. The spectator's job is to guess when the performer lies. This is carried through, one card at a time. After each time, the performer shows the actual card and enters a score of correct or incorrect. This is continued for ten or so steps. At the end, the computer gives a tally of, say, "Seven correct out of ten, above average."

Now the tone changes. It is the spectator's turn to lie or tell the truth. What's more, it won't be the performer who guesses "lie or truth." The computer will act as a lie detector. The spectator looks at the top card of the pack and decides (mentally) whether to lie or not. Depending on which decision he or she makes, the spectator taps the "R" or "B" key on the computer to indicate red or black. The computer responds, accurately determining if a lie is told. The messages vary from time to time but the computer is always right. This has an eerie effect, quite out of proportion to the trick's humble means.

How does it work? The deck is set up initially with red/black alternating throughout. The spectator cuts the deck several times and deals some cards into a single pile on the table. The performer might patter about poker and bluffing or lie detector machines. The two piles are riffle shuffled together by the spectator, who then deals them into two piles, alternating left, right, left, right, and so on, until ten or so cards are in each pile. The spectator hands either pile to the performer.

Here is the key to the trick. Because of the Gilbreath Principle, each consecutive pair of cards contains one red and one black after the riffle shuffle. Dealing alternately into two piles ensures that the cards are of opposite colors in the two piles as we work from top to bottom. Thus, if the top twenty cards of the deck after shuffling are RBBRBRRBRB-BRRBBRRBBR, then after dealing into two alternate piles, we have

R	B
B	R
B	R
R	B
R	B
B	R
R	B
B	R
R	B
B	R

If the top of the left-hand pile is red, the top of the right-hand pile is black. The same holds for the second cards, and so on. The spectator hands either packet to the performer who looks at the cards, calls out colors, and lies or not each time. There is no preset pattern. Just do as you please, using funny tones of voice and making faces if that's your style. The spectator guesses "lie or truth," the performer shows the card, and enters "C" or "W" each time, depending on whether the guess is correct or wrong.

The second secret lies here. After each guess is entered on the computer, the performer taps the space bar if the actual card in question is red, and does not tap if the actual card is black. This variation goes unnoticed amidst all the banter, and it tells the computer the actual colors of the cards in the performer's pile. By taking opposites, the computer now knows the actual color of each card in the spectator's pile. When the spectator goes through his or her pile (and whatever complex thought processes are required), he or she finally presses the "R" or "B" key. The computer compares each of the spectator's entries with the known color and determines if a lie has been told.

It will help the presentation if a separate set of messages is preprogrammed for each card. Thus, the computer might announce, "You lie" or "Tsk-tsk—don't try that again" for lies, or "You're trying to trick me—you told the truth" when the spectator isn't lying. This takes a modest amount of preparation but is worth the effort.

When Paul Curry first performed this for us, personal computers and programmable calculators were far in the future. He hand-built a complicated gadget with displays, wires, and switches all over it to carry out this simple task. He later published a pencil and paper version of the trick in his wonderful book *Paul Curry Presents*. Because this loses the wonderful effect of the computer as lie detector, it is not as good as the version above.

We will not give programming details here. If you know a bit about programming, it's an hour's work (oh, all right, a few hours' work). If you don't, go find a teenager. The Curry trick is a great example of how thought and presentation can turn a humble mathematical trick into great theater. Curry also invented perhaps the greatest red/black trick of all time: Out of This World. We can't explain it here but it is definitely worth hunting down.

So far we have explained Gilbreath's First Principle. In 1966, Gilbreath stunned the magical world by introducing a sweeping generalization, known as Gilbreath's Second Principle. In the first principle, alternating red/black patterns are used. Gilbreath discovered that *any* repeated pattern can be used. For example (go get a deck of cards), arrange a normal deck so that the suits rotate: clubs, hearts, spades, diamonds, clubs, hearts, spades, diamonds, and so on. Give the deck a random cut, deal any number of cards onto the table face-down in a pile (reversing their order), and riffle shuffle the two piles together. The top four cards will consist of one of each suit, no repeats, the next four cards will have one of each suit, and so on through to the bottom four cards in the deck.

Here is a simple variation. Remove all four aces, twos, threes, fours, and fives from the deck (twenty cards in all). Arrange them in rotation:

$$1, 2, 3, 4, 5, 1, 2, 3, 4, 5, 1, 2, 3, 4, 5, 1, 2, 3, 4, 5.$$

Cut this packet randomly, deal any number of cards face-down into a pile on the table, and riffle shuffle this pile with the rest of the cards from the packet. The top five cards will be {1, 2, 3, 4, 5} in some order, the same for the next five, the next five, and the last five cards. Dozens of tricks have been invented using these ideas. Often this principle is combined with some sleight of hand, making the trick unsuitable for this book. We have cleaned up one of them to make quite a performable trick. Those knowing a bit of sleight of hand will be able to dress it more handsomely (see chapter 11 if you want to learn more).

The following has served us well. It uses Gilbreath's Second Principle together with ideas from Ronald Wohl and Herbert Zarrow.

The rough effect is this: The performer asks if someone would like a lesson in cheating at cards. "A key to making big money is that you must learn to deal someone a good hand but also deal a better hand to yourself (or your partner)." With these preliminaries, the spectator cuts, shuffles, and deals the cards (with a little help and kidding from the performer). The spectator deals a pat poker hand—a high straight (ace, king, queen, jack, ten), to one player and a better hand, a flush (all five cards of the same suit), to himself. At the end, the spectator is as mystified as everyone else. The performer cautions that the new skills are to be used for entertainment purposes only.

To perform this trick, the top twenty-five cards of the deck must be prearranged. Remove any ten spades and three each of aces, kings, queens, jacks, and tens (of any suits). These are arranged with five spades on the top, five spades on the bottom, and the middle fifteen cards in the rotation ace, king, queen, jack, ten, and so on, that is, as:

S S S S S A K Q J 10 A K Q J 10 A K Q J 10 S S S S S.

Put these twenty-five cards on top of the rest of the cards and put the cards in the card case.

Ask for a volunteer who wants to learn about cheating at cards. This may involve some funny interactions with the audience. Ask the volunteer (let's call her Susan) if she knows how to play poker—with all the poker on TV, many people do. But still, many people don't. Take out the deck, turn it face-up and display a few poker hands, explaining one pair, two pairs, three of a kind, straight, and flush. Do this without disturbing the original top twenty-five cards. Saying you'll start easy, break the deck (it's still face-up) at the run of five middle spades, so you have only the original top twenty-five in hand. Turn these face-down and hand them to Susan. Say you're going to get an idea of her dealing skills—ask her to deal any number of cards into a single pile on the table. The actual number doesn't matter but it must be five or more, and at most twenty. Now ask if she can shuffle, and have her riffle shuffle the dealt pile with the rest of the twenty-five cards. Tell her that poker is played by dealing around—have her deal five hands as in a normal poker game. Comment on her technique. Have her look at one of the hands (turn it face-up, without changing the order, and comment on its value). Now have her assemble the five hands in any order, keeping the packets of five together.

Say, "That was practice; here comes the real thing. There is a high roller in second position and your partner is playing the first hand. I'd like you to use your skills to deal a pat hand to the second player but make sure you give your partner a better hand." Susan may look at you as if you are out of your mind. Anyway, have her deal five hands in the normal fashion. Turn up the cards in second position one at a time. They will form an ace-high straight—shake her hand, and act as if the trick is over. "Susan, you're really talented." Then remember, "Wait. An ace-high straight is almost impossible to beat. The odds of getting an ace-high straight are about one in twenty-five hundred. What about your partner?" Slowly turn over the cards in the first player's hand one at a time. They will form a flush in spades, handily beating the ace-high straight. Offer her your hand again with the comment, "Susan, you're a poker genius."

That's a lot of dressing but it makes for a very entertaining few minutes. For you, our reader, understanding how it all comes together is a nice lesson in the beginnings of combinatorics. A fancier version of this trick involving some sleight of hand appears as U-shuffle Poker in *Zarrow, A Lifetime of Magic* by David Ben.[2]

THE ULTIMATE GILBREATH PRINCIPLE

Up to now, we have seen two grand applications of Gilbreath's two principles, utilizing reds and blacks as well as rotating sequences. It is natural to ask what other properties or arrangements are preserved by our riffle shuffle. This is actually a hard, abstract math question. What do we mean by "property or arrangement" and "preserved"? After all, if a deck of cards labelled {1, 2, 3, . . . , 52} is shuffled in any way, it still contains all these numbers only once. Clearly, this doesn't count. Is looking at every other card allowed?

Let us start by carefully defining what we mean by "shuffle." Consider a deck of N cards labeled 1, 2, 3, . . . , N. A normal deck has $N =$ 52. The deck starts out in order, with card 1 on top, card 2 next, and card N on the bottom. By a *Gilbreath shuffle* we mean the following permutation. Fix a number between 1 and N, call it j. Deal the top j cards into a pile face-down on the table, reversing their order. Now, riffle shuffle the j cards with the remaining $N - j$ cards. For example, if $N =$ 10 and $j = 4$, the shuffle might result in:

1				4	
2				5	
3	5			6	
4	6	4		3	
5	→	7	3	→	7
6	8	2		2	
7	9	1		8	
8	10			9	
9				1	
10				10	

What we want to understand is just what arrangements are possible after one Gilbreath shuffle? Two answers will be given. First, we will count how many different arrangements are possible. Second, we will give a simple description of the possible arrangements, which we modestly call the Ultimate Gilbreath Principle.

COUNTING. The number of different permutations of N cards is $N \times (N - 1) \times (N - 2) \times \cdots \times 2 \times 1 = N!$ (read "N factorial"). These numbers grow very rapidly with N. For example, if $N = 10$ then $N!$ = 3,628,800, more than three and a half million. When $N = 60$, $N!$ is larger than the number of atoms in the universe. Put another way, $60! \approx 8.32 \times 10^{81}$, while the estimated number of atoms in the universe (according to the current theories) is less than 10^{81}.

Of course, after one Gilbreath shuffle, not all arrangements are possible. At the end of this chapter we show that, with a deck of N cards, only 2^{N-1} arrangements can occur. When $N = 10$, $2^{N-1} = 512$. When $N = 52$, $2^{51} \approx 2.25 \times 10^{15}$. This is still a large number (which makes the tricks confusing and interesting). As an example, the reader may check that, with four cards, the eight possible Gilbreath arrangements are:

```
1  2  2  2  3  3  3  4
2  1  3  3  2  2  4  3
3  3  1  4  1  4  2  2
4  4  4  1  4  1  1  1.
```

As an aside, we first did the examples for decks of sizes $N = 1, 2, 3, 4$. By enumerating all possibilities by hand, we saw the answer 2^{N-1}. If this is the right answer, it is so neat that there must be an easy proof. Notice that having a neat count is different from having a neat description. We give our descriptions next, followed by an appeal for help in inventing a good trick. The proofs are given at the end of the chapter.

THE ULTIMATE INVARIANT(S)

To describe the results, we need some way of writing things down. For a deck of cards originally in order $1, 2, 3, \ldots, N$, record a new order (we call it π) by letting $\pi(1)$ be the card at position 1, $\pi(2)$ be the card at position 2, $\ldots$, and $\pi(N)$ be the card at position N. Thus, if the new order of a five-card deck is 3, 5, 1, 2, and 4, then $\pi(1) = 3$, $\pi(2) = 5$, $\pi(3) = 1$, $\pi(4) = 2$ and $\pi(5) = 4$. This may seem like a complex way to talk about something simple, but we can't proceed without it. Thus, we can now say that "π is a Gilbreath permutation" is shorthand for "A deck of N cards starting in order $1, 2, 3, \ldots, N$ is in final order $\pi(1)$, $\pi(2), \ldots, \pi(N)$ after one Gilbreath shuffle."

The final thing we need is the notion of the remainder modulo j. If we take a fixed number j (e.g., $j = 3$), then any number (e.g., 17) has some remainder when divided by j. For example, 17 has remainder 2 when divided by 3. In this case, we say 17 is 2 modulo 3. A set of numbers are *distinct modulo j* if their remainders are distinct. Thus, 12 and 17 have remainders 0 and 2 and so are distinct modulo 3, whereas

14 and 17 are not, since 14 and 17 both have the same remainder 2 modulo 3. With all these prerequisites, here is our main result. The abstract-looking statement is followed by some very concrete examples. The proof is given below.

> THEOREM. The Ultimate Gilbreath Principle. For a permutation π of $\{1, 2, 3, \ldots, N\}$, the following four properties are equivalent:
> 1. π is a Gilbreath permutation.
> 2. For each j, the top j cards $\{\pi(1), \pi(2), \pi(3), \ldots, \pi(j)\}$ are distinct modulo j.
> 3. For each j and k with $kj \leq N$, the j cards $\{\pi((k-1)j+1), \pi((k-1)j+2), \ldots, \pi(kj)\}$ are distinct modulo j.
> 4. For each j, the top j cards are consecutive in $1, 2, 3, \ldots, N$.

Here is an example illustrating the theorem. For a ten-card deck, we can deal off four cards into a small pile on the table (one by one) and then riffle shuffle them to lead to the arrangement π below:

$$4$$
$$5$$
$$6$$
$$3$$
$$7$$
$$2$$
$$8$$
$$9$$
$$1$$
$$10.$$

Thus, π is a Gilbreath permutation, so it satisfies (1) by definition. The theorem now says that π has many special properties. For example, consider property (2). For each choice of j, the remainders modulo j of the top j cards are distinct. When $j = 2$, the top two cards, 4 and 5, have distinct remainders 0 and 1 modulo 2. When $j = 3$, the top three cards, 4, 5, and 6, are 1, 2, 0 modulo 3. This works for all j up to N, no matter what Gilbreath shuffle is performed.

Property (3) is our refinement of the original general Gilbreath Principle. For example, if $j = 2$, it says that, after any Gilbreath shuffle, each consecutive pair of cards contains one even value and one odd value. If the even cards are red and the odd cards are black in the original arrangement, we have Gilbreath's First Principle. The small refinement is that we do not need to assume that N is divisible by j; the last k cards still have distinct remainders when divided by j, provided $k \leq j$ and the number of cards preceding these cards is a multiple of j.

The final one, property (4), needs some explanation. Consider our Gilbreath permutation π (written sideways to conserve space):

$$4\ 5\ 6\ 3\ 7\ 2\ 8\ 9\ 1\ 10.$$

The top four cards (here 4 5 6 3) were consecutive in the original deck. (They are out of order, but the set of four started out as consecutive). Similarly, for any j the top j cards were consecutive in the original deck for any j.

The point of all of this is that *any one of these* parts gives a complete characterization. For example, if π is a Gilbreath permutation then π satisfies property (3) for all j. Conversely, if π is any permutation satisfying property (3) for all j, then π arises from a Gilbreath shuffle. In one sense, this is a negative result. It says that there are no new hidden invariants—Gilbreath discovered them all. On the other hand, now we know and can stop brooding about this.

Property (2) is our new Ultimate Gilbreath Principle. We haven't seen it elsewhere and it is the key to proving the theorem. What we don't see is any way of making a good trick. In the hope of angering some readers into making progress in this direction, here is an unsuccessful attempt.

You, the performer, show ten cards, each with a unique number, 1, 2, 3, . . . , 10. The patter goes as follows: "Did you ever have to help your kids with their math homework? It's getting pretty complicated. Our kids are doing binary, ternary, and octal arithmetic. They came home with something they call 'modulo j.'" Explain modulo j (as we did before) and then continue, "Their teacher says the following stunt always works." The cards are arranged in order, say, 1, 2, . . . , 10. They can be ordinary playing cards or index cards with bold numbers written on them. Have the spectator cut the packet, deal any number onto the table, and then shuffle the two packets together. Explain: "The top two cards are a full set modulo 2, so one should be even and the other odd. Let's take a look. Now, the top three cards form a complete set modulo 3. Turn over the next card." Explain how it's true: "Let's see—4, 5, 3, well 3 is 0 modulo 3, and 4 is 1 modulo 3 and 5 is 2 modulo 3, so it worked then. Let's see the next card. . . ." This continues for as long as you have the nerve to keep talking.

To be honest again, we haven't had the courage to try this trick out on our friends. It just doesn't seem very good. What's worse, the pattern described in property (4) of the theorem might be obvious. Indeed, this is the way (4) was discovered. We had proved the equivalence of (1), (2), and (3) without knowing (4). When we tried the trick out, we noticed (4). Its discovery makes the proof much easier. Take a look at the proof in the following section.

THE MANDELBROT SET

The Mandelbrot set is one of the most amazing objects of mathematics. Figure 2 shows a picture of the Mandelbrot set. A close look reveals a "leafy" quality on the edge of everything. Consider the bottom region of figure 2. We blow this area up in figure 3. Now, new "leafy" fixtures appear. The bottom region of figure 3 is expanded to reveal the dazzling structure in figure 4. Figures 5 and 6 take closer and closer looks. Each reveals a rich, detailed structure.

There are many computer programs on the Web that allow exploration of the Mandelbrot set.[3] The appearance of refined structure keeps going forever. It has engaged the best minds in mathematics, physics, and biology. Moreover, as explained below, the pattern is "universal." It appears in many other seemingly unrelated systems.

This is a chapter on shuffling cards and the Gilbreath Principle. We hope the reader is as surprised as we were to learn that there is an intimate connection between shuffling cards and the Mandelbrot set. The story is hard to tell, so here is a roadmap to what's coming. We begin with a simple procedure: squaring and adding. This is really all that is needed to define the Mandelbrot set. Next, we determine when repeated squaring and adding leads to a periodic sequence. Card shuffling and the Gilbreath Principle now enter to describe the way the points of this sequence are ordered. All of the activity up to now has taken place with one-dimensional, "ordinary" numbers. The Mandelbrot set lives in *two* dimensions. Only then can the Mandelbrot set be properly defined. At the end, we give a whirlwind tour of the Mandelbrot set, explain its universality, and enter a plea for help in finding two-dimensional shuffles that will explain the last remaining mysteries.

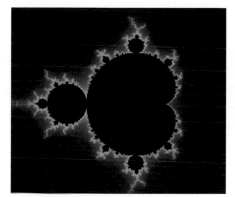

Figure 2. The Mandelbrot set (image created by Paul Neave, neave.com)

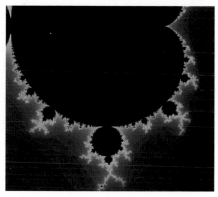

Figure 3. Enlarging part of the Mandelbrot set (image created by Paul Neave, neave.com)

Figure 4. A further enlargement of part of the Mandelbrot set (image created by Paul Neave, neave.com)

Figure 5. A further enlargement (image created by Paul Neave, neave.com)

Figure 6. An even further enlargement (image created by Paul Neave, neave.com)

SQUARING AND ADDING. Repeated squaring is a familiar procedure. Starting with 2, we get 2, 4, 16, 256, 65536, . . . , off to infinity. Starting with a number less than 1, say with $1/2$, we get $1/2$, $1/4$, $1/16$, $1/256$, $1/65536$, ... This sequence tends to zero. We will have to deal with negative numbers. Starting at -1 and repeatedly squaring gives -1, 1, 1, 1, 1, 1, ... Things become more interesting if a fixed number is added each time after. Suppose 1 is added each time. Starting with 0, squaring and adding 1 gives $0^2 + 1 = 1$, squaring and adding 1 repeatedly gives $1^2 + 1 = 2$, $2^2 + 1 = 5$, $5^2 + 1 = 26$, . . . off to infinity. If instead we add -1 each time, we get 0, $0^2 - 1 = -1$, $(-1)^2 - 1 = 0$, $0^2 - 1 = -1$, $(-1)^2 - 1 = 0$, ... This sequence bounces back and forth between 0 and –1 forever. It's the same if we add –2 each time. This time the sequence goes 0, –2, 2, 2, 2, 2, ... Adding any number smaller than –2 or larger than 0 leads to a sequence that tends to infinity. Starting points between –2 and 0 lead to bounded sequences. (They don't get arbitrarily far from 0 as time goes on). They are in the Mandelbrot set.

PERIODIC POINTS. Adding certain special numbers leads to sequences that cycle in a fixed pattern. Let c be the value added after each squaring. Thus, the sequences are:

$$0, \, 0^2 + c = c, \, c^2 + c, \, (c^2 + c)^2 + c = c^4 + 2c^3 + c^2 + c, \ldots$$

If such a sequence is to return to 0, then eventually one of the iterated terms must vanish. Consider the term $c^2 + c$. When is this 0? If $c^2 + c = 0$ then either $c = 0$ or $c + 1 = 0$, i.e., $c = -1$. We saw above that adding –1 each time gives 0, –1, 0, –1, 0, –1, . . . , a pattern with "period 2." Consider the next term $c^4 + 2c^3 + c^2 + c$. Which values of c make this 0? The value $c = 0$ works but we have seen this before. If $c \neq 0$, we can divide through and consider $c^3 + 2c^2 + c + 1$. This is a cubic equation and there is a rather complicated formula for the roots of a cubic polynomial that shows that in this case, the value

$$c = -\frac{\sqrt[3]{100 + 12\sqrt{69}}}{6} - \frac{2}{\sqrt[3]{100 + 12\sqrt{69}}} - \frac{2}{3} = -1.75487\ldots$$

works. Using this value for c, we get

$$0, \, -1.75487\ldots, \, (-1.75487\ldots)^2 - 1.75487\ldots = 1.32471\ldots, \, 0, \ldots$$

This pattern continues, repeating every third step. We say that $c = -1.75487\ldots$ is a "period three" point.

The same scheme works to get points of a higher period. For example, squaring $c^4 + 2c^3 + c^2 + c$ and adding c gives $c^8 + 4c^7 + 6c^6 + 6c^5 + 5c^4 + 2c^3 + c^2 + c$. This gives two new values of c, both of which lead to points of period four. These are $c = -1.3107\ldots$ and $c = -1.9407\ldots$. These in turn lead to the repeated sequences

$$c = -1.3107\ldots: \quad 0, \, -1.3107\ldots, \, 0.4072\ldots, \, -1.1448\ldots, \, 0, \ldots$$

and

$$c = -1.9407\ldots: \quad 0, \, -1.9407\ldots, \, 1.8259\ldots, \, 1.3931\ldots, \, 0, \ldots .$$

New periodic sequences occur for each possible period. These can be found by finding the values of c where the nth iterate of "squaring and adding c" vanishes. They turn out to be exactly described by Gilbreath permutations.

THE SHUFFLING CONNECTION. To make the connection with shuffling cards, write down a periodic sequence starting at zero. Write a one above the smallest point, a two above the next smallest point and so on. For example, if $c = -1.75487\ldots$ (a period three point), we have:

$$\begin{array}{cccc} 2 & 1 & & 3 \\ \hline 0 & -1.75487\ldots & 1.32471\ldots \end{array}$$

For the two period four sequences, we get for $c = -1.3107\ldots$:

$$\begin{array}{cccc} 3 & 1 & 4 & 2 \\ \hline 0 & -1.3107\ldots & 0.4072\ldots & -1.1448\ldots \end{array}$$

and for $c = -1.9407\ldots$:

$$\begin{array}{cccc} 2 & 1 & 4 & 3 \\ \hline 0 & -1.9407\ldots & 1.8259\ldots & 1.3931\ldots \end{array}$$

For a fixed value of c, the numbers written on top code up a permutation that is a Gilbreath shuffle. Here is the decoding operation. For example, when $c = -1.3107\ldots$, the numbers on top are 3 1 4 2. Start with the 1 and go to the left (going around the corner if you have to). This gives (1324). This is "cycle notation" for a permutation. It is read as "1 goes to 3, 3 goes to 2, 2 goes to 4, and 4 goes back to 1." Rewrite this by putting the numbers 1, 2, 3, 4 in a row, and under them put what they go to in the cycle, as:

$$\begin{array}{cccc} 1 & 2 & 3 & 4 \\ 3 & 4 & 2 & 1. \end{array}$$

The reader may practice by taking the example $c = -1.9407\ldots$. As we have seen above, it is 2, 1, 4, 3. Starting with 1 and going to the left gives the cycle (1234), and finally the two-line arrangement

$$\begin{array}{cccc} 1 & 2 & 3 & 4 \\ 2 & 3 & 4 & 1. \end{array}$$

The point of all this decoding is that the arrangement on the bottom line is always a Gilbreath permutation, and furthermore, *every* cyclic Gilbreath permutation of length n appears exactly once from a period n value of c.

We were told this result by Dennis Sullivan, who attributes it to the great mathematicians John Milnor and William Thurston. These are three of the greatest mathematicians of the twentieth century—the latter two are winners of the Fields Medal (often called "the Nobel Prize of mathematics"). Whomever this result belongs to, it sets up a fascinating connection that is just beginning to be understood. We state it as a formal theorem below along with further comments.

THE FULL MANDELBROT SET AT LAST. All of the activity above has been confined to the one-dimensional line. The Mandelbrot set lives in two dimensions. There is a notion of "square and add c" in two dimensions. The values of c where repeated squaring and adding stays bounded are exactly the points of the Mandelbrot set. Working in the plane, the values of c are two-dimensional: $c = (c_1, c_2)$.

Figure 1 shows the Mandelbrot set. The values *on* the x-axis between -2 and 0 are the points discussed above. The big central heart-shaped region is called the cardioid. It is surrounded by blobs, and each of these is in turn surrounded by smaller blobs (and so ad infinitum). One of the main open research problems concerning the Mandelbrot set has to do with the values of c (now (c_1, c_2)) that give periodic sequences from the squaring and adding process. It is conjectured that each blob (the big ones, the smaller ones, and so ad infinitum) contains one of those periodic points. Proving this conjecture would lead to the resolution of the outstanding *local connectivity* conjecture. Sullivan told us about the connection with shuffling because shuffles parameterize the periodic points on the x-axis. Is there some kind of two-dimensional shuffle that parameterizes the two-dimensional periodic points? We don't know but we're thinking hard about it.

☠ SOME MATH WITH A BIT OF MAGIC ☠

Squaring and adding makes perfect sense in two dimensions, taking a point z to $z^2 + c$. There is a simple geometric description: A point z in two dimensions is described by its coordinates (x, y). Figure 7 shows (x, y) plotted as a dot with the line connecting the dot to 0. Also shown is the angle θ that the point (x, y) makes with the x-axis. To square the

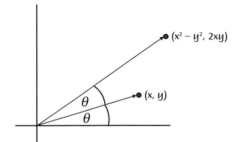

Figure 7. Squaring a complex point

point (x, y), we square the distance along the line connecting (x, y) to 0, double the angle to 2θ, and plot the new point. The new point can be given in coordinates as $(x^2 - y^2, 2xy)$. Call this (x', y'). Adding $c = (c_1, c_2)$ gives $(x' + c_1, y' + c_2)$. This is repeated using the same value of c each time. If this procedure, starting at 0, leads to points that stay inside a large enough circle around 0, we put c in the Mandelbrot set. Figure 2 shows all such values of c.

A comprehensive picture book about the Mandelbrot set is *Chaos and Fractals: New Frontiers of Science*, by H. O. Peitgen, H. Jürgens, and D. Saupe.[4] A discussion of shuffling and the Mandelbrot set can be found in the paper "Bounds, Quadratic Differentials, and Renormalization Conjectures" by D. Sullivan.[5] To observe professionals talking among themselves about the Mandelbrot set, see T. Lei's book *The Mandelbrot Set: Theme and Variations*.[6]

Let us state the basic connection between shuffling cards and real points in the Mandelbrot set more carefully. Define a sequence of polynomials $P_1, P_2, P_3, \ldots$, iteratively as $P_1(x) = x$, $P_2(x) = x^2 + x$, $P_3(x) = P_2(x)^2 + x = (x^2 + x)^2 + x = x^4 + 2x^3 + x^2 + x, \ldots, P_n(x) = P_{n-1}(x)^2 + x$. Thus, the top degree of P_n is 2^n. Dennis Sullivan proved that the real zeroes of P_n are "simple." Each real zero can be used as the additive constant in the "squaring and adding" iteration.

> THEOREM. Define $P_1(x) = x$ and $P_{k+1} = P_k^2 + x$ for $h < n$. The real zeroes c of P_n that lead to periodic sequences of period n are in one-to-one correspondence with Gilbreath permutations that are n-cycles. The correspondence goes as follows: From c, form the iteration $0, c, c^2 + c, c^4 + 2c^3 + c^2 + c, \ldots$. Label the smallest of those values 1, the next smallest 2, $\ldots$, and the largest n. Read these in right-to-left order as a cyclic permutation. Transform this to two-rowed notation. The resulting bottom row is a Gilbreath permutation (characterized at the beginning of this chapter). Each cyclic Gilbreath permutation occurs exactly once through this correspondence.

Note that not every Gilbreath permutation gives rise to an n-cycle. For example, removing the top card and inserting it into the middle of the deck is a Gilbreath permutation that is not an n-cycle. The number of n-cycles among Gilbreath permutations has been determined by Rogers and Weiss.[7] They show that this number is exactly

$$\frac{1}{2n} \sum_{d \mid n, d\, odd} \mu(d)\, 2^{n/d}.$$

Here, the sum is over the odd divisors d of n, and $\mu(d)$ is the so-called Möbius function of elementary number theory. That is, $\mu(d)$ is 0 if d is divisible by a perfect square, and $\mu(d) = (-1)^k$ if d is the product of k distinct prime factors. Thus, for $n = 2, 3, 4, 5, 6$, the formula gives $\frac{1}{4}(2^2) = 1$, $\frac{1}{6}(2^3 - 2) = 1$, $\frac{1}{8}(2^4) = 2$, $\frac{1}{10}(2^5 - 2) = 3$, and $\frac{1}{12}(2^6 - 2^2) = 5$ cyclic Gilbreath permutations. For example, the three values of c for $n = 5$ give:

2	1	5	4	3
0	−1.9854	1.9564	1.8424	1.40900

3	1	5	4	2
0	−1.8607	1.6017	0.7047	−1.3640

4	1	5	3	2
0	−1.6254	1.0165	−0.5920	−1.2749

These lead, respectively, to the two-line arrays:

1	2	3	4	5
2	3	4	5	1

1	2	3	4	5
3	4	2	5	1

1	2	3	4	5
4	3	5	2	1

where the second rows are Gilbreath permutations.

Recall that there are exactly 2^{n-1} Gilbreath shuffles. The formula above shows that there are approximately $\frac{2^{n-1}}{n}$ Gilbreath n-cycles. Jason Fulman gives a formula for the number of unimodal permutations with a given cycle structure.[8]

To conclude, let us try to explain in what sense the Mandelbrot set is universal. For fixed c, the square and add operation changes x to $x^2 + c$. As c varies, we have a family of different iteration schemes. Curt McMullen showed that *any* family of functions of the plane to itself has all the complexity of the Mandelbrot set complete with its holes, fractal dimensions, and infinite subtlety. Of course, this also means it contains all the Gilbreath permutations described above. A more careful version of McMullen's theorem strains the confines of this page.[9]

We know of two applications of Gilbreath's principles outside of magic. The mathematician N. G. de Bruijn (whom we met in chapters 2, 3, and 4) published "A Riffle Shuffle Card Trick and Its Relation to Quasicrystal Theory" in 1987.[10] The quasicrystals referred to are Penrose tiles. These are two shapes of tiles that can be used to tile the plane but only in a nonperiodic way (see figures 8 and 9). They have a fascinating story, which is detailed in Marjorie Senechal's book *Quasicrystals and Geometry*, or the more accessible *Miles of Tiles*, written by the mathematician Charles Radin. Most accessible of all is Martin Gardner's treatment in *Penrose Tiles to Trapdoor Ciphers*.[11]

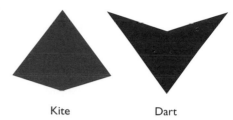

Kite   Dart

Figure 8. The two Penrose tiles

A BIT OF MAGIC

De Bruijn shows that the Gilbreath Principle leads to understanding useful facts about the properties of Penrose tilings. Along the way, de Bruijn worked the following extension of Gilbreath's First Principle. Before starting, separate the cards so you have all clubs together, hearts together, spades together, and diamonds together. Form one twenty-six-card pile with spades and diamonds alternating (SDSD . . .). Then form another twenty-six-card pile with hearts and clubs alternating (HCHC . . .). If the two piles are riffle shuffled together, we know from before that each consecutive pair will consist of one red and one black. However, if the two piles are put together and the deck of fifty-two is cut freely, this need not work out. De Bruijn's "extension" allows a free cut. He proves that *either* each consecutive pair contains one red and one black throughout, *or* each consecutive pair contains one major suit (i.e., a heart or spade) and one minor suit (i.e., a club or diamond). With suitable arrangement, major/minor may be replaced by odd/even or high/low, which might be more suited to a magic trick.

Figure 9. Tiling with Penrose tiles

De Bruijn's extension goes beyond the original Gilbreath. In light of our theorem, how can this be? De Bruijn adds an extra restriction (the packet cut off is not of a freely chosen size), but he gets his freedom at the end—a free cut. We have tried to marry de Bruijn's extension with our ultimate principle. The mix makes a nice example of how progress occurs.

The second application outside card magic comes in the world of designing sorting algorithms for computers. Huge memory files are often stored on external discs. Several discs can be read at once.

Stanford computer scientist Donald Knuth used the Gilbreath Principle to give an "improved superblock striping" technique that allows two or more files, distributed on discs, to be merged without possible conflict (in other words, the need to read two blocks from the same disc at the same time). This is explained in Knuth's monumental book series *The Art of Computer Programming*. A videotape of Knuth's talk on this superblock striping technique is available from the Stanford University Computer Science Department.

SOME PROOFS. We continue by providing the promised proofs for some of the theorems above. Why are there 2^{N-1} Gilbreath shuffles of N cards? Let us select some arbitrary subset $S = \{s_1, s_2, \ldots, s_j\}$ of $\{2, 3, \ldots, N\}$. Now form the Gilbreath permutation by placing j in position 1, then $j-1$ in position s_1, $j-2$ in position s_2, etc., and placing the numbers greater than j in *increasing order* in the positions *not* in S. It is clear that all Gilbreath permutations can be uniquely built this way. Since the number of ways of choosing S is just 2^{N-1}, then we have the desired result.

Here is a proof that the four properties (1), (2), (3), and (4) listed in our Ultimate Gilbreath theorem are all equivalent. The arguments are elementary but not so easy to discover. They make nice examples of how card tricks can lead to mathematics.

PROOF. After a Gilbreath shuffle, the top j cards form an interval $\{a, a+1, \ldots a+j-1\}$ or $\{a, a-1, \ldots, a-(j-1)\}$ for some value of a. As such, they consist of distinct values modulo j. Thus, (1) implies (2). If π satisfies (2) for each j then π satisfies (3). To see this, consider π satisfying (2). Clearly, the entries in the first block are distinct. But the top $2j$ are also distinct modulo $2j$ and consist of exactly two of each value modulo j. Since the top j are one of each value modulo j, it must be that $\pi(j+1), \pi(j+2), \ldots, \pi(2j)$ are distinct modulo j. This in turn implies that $\pi(2j+1), \pi(2j+2), \ldots, \pi(3j)$ are distinct, and so on. Clearly, (3) implies (2), so (2) and (3) are equivalent.

To see that (2) implies (1), observe that (2) implies that the top j cards form an interval of values. Suppose the top card $(\pi(1))$ is k. The next card must then be $k+1$ or $k-1$, since if it is $k \pm d$ for some $d > 1$, then the top d cards would not be distinct modulo d. Suppose the top $j+1$ cards were $a, a+1, \ldots, a+j$. If the next was not $a-1$ or $a+j+1$, but $a+j+d$ for some $d>1$, then, again, modulo d, things would repeat.

Finally, this "interval" property of π implies it can be decomposed into two chains $k+1$, $k+2$, . . . , n and k, $k-1$, . . . , 1. For this, proceed sequentially. If the top card is k, the next must be $k+1$ or $k-1$. Each value that increases the top of the interval is put in one chain, and each value that decreases the bottom is put in the second chain. Since, for such intervals, increasing values occur further down in π, the two chains formed do the job. This finishes the proof (whew!).

FURTHER REMARKS

1. The decomposition into two chains is not unique. If we deal off k cards and, in the shuffle, $k+1$ is left above k, it is impossible to distinguish this from $k+1$ being dealt off.

2. Instead of dealing, we can cut off and turn a packet of k face-up, then shuffle the two packets together.

3. As a last mathematical detail: At the start of this chapter we gave a heuristic calculation of the chance that a well-shuffled deck of $2N$ cards has one red and one black card in each consecutive pair. Naïve heuristics suggest that when N is large, the pair choices are roughly independent and each one has the $\frac{1}{2}$ chance of coming up red/black in some order. This would result in a probability of $\frac{1}{2^N}$ of happening. However, the events we are considering are *not* independent. In particular, if we start with a shuffled deck of N red and N black cards, the chance that, after the first card is selected, the *next* card selected has a different color from the first card is slightly *greater* than $\frac{1}{2}$. After all, there are only $N-1$ cards with the first card's color left in the deck, while there are still N cards with the opposite color. This imbalance happens for each of the pairs selected, and becomes greater as the number of cards gets smaller. For example, for a four-card deck (i.e., $N=2$), the chance that the first two cards form a red/black pair is $\frac{2}{3}$. The result of multiplying all these "imbalances" together is that the probability that our well-shuffled deck will have the desired property is exactly $\frac{2^N}{\binom{2N}{N}}$, which is approximately equal to $\frac{\sqrt{\pi N}}{2^N}$, using the Stirling approximation again. For $N=26$, this is $1.353\ldots \times 10^{-7}$, which amounts to less than one chance in seven million.

SOME HISTORY

Gilbreath's First Principle originally appeared as the trick Magnetic Colors in the magic magazine the *Linking Ring* in July 1958. The *Linking Ring* is the official publication of one of the two largest American magic organizations, the International Brotherhood of Magicians, or IBM. (The other is the Society of American Magicians, or SAM.) The *Linking Ring* has been published monthly since 1923. A typical issue contains advertisements from magic dealers, historical articles, editorials denouncing magical exposés, and a large section of tricks contributed by IBM members. You cannot find it in libraries. As with most magical information, it is for magicians only.

Back in 1958, young Norman Gilbreath introduced himself in the magazine as follows: "I have been interested in magic for 10 years. I am a math major at the University of California in Los Angeles (UCLA). Being a supporter of the art of magic, I have created over 150 good tricks and many others not so good. Here are a couple I hope you can use." He then provided a brief description of what is now called Gilbreath's First Principle, in which he dealt the deck into two piles, following the shuffle, and revealed that the cards in each pair have opposite colors.

Gilbreath's trick was picked up and varied almost immediately. In the January 1959 issue of the *Linking Ring*, card experts Charles Hudson and Edward Marlo wrote, "It is not often one runs across a new principle in card magic. . . . Norman Gilbreath's 'Magnetic Colors' has proven the most popular card effect to appear in the parade for a long time." Gilbreath weighed in eight years later by introducing his second principle in the June 1966 issue of the *Linking Ring*. By this time, Gilbreath was a professional mathematician working for the Rand Corporation. He held this job for his entire career. The second principle was featured in this special issue of the magazine devoted to Gilbreath's magic. It included new uses for the first principle and many noncard tricks. Gilbreath published later variations that involved mixing red decks with blue decks and face-up cards with face-down cards (with some effort, you can find these in the magic magazine *Genii*).[13]

The nonmagical public heard about the Gilbreath Principle in Martin Gardner's *Scientific American* column in August 1960. He expanded this into a chapter in his third book, *New Mathematical Diversions from "Scientific American."* New presentations and applications have regularly appeared in magic journals. A booklet titled "Gilbreath's Principles," written by mathematics teacher and magician Reinhard Muller, appeared in 1979. Chapter 6 of Justin Branch's *Cards in Confidence*, vol. 1, is filled with many variations.[14] While our Ultimate Gilbreath Principle shows there can be no really new *principle*, the variations make for good magic.

Chapter 6

NEAT SHUFFLES

Some magicians, and some crooked gamblers, can shuffle cards perfectly. This means cutting the deck exactly in half and riffling the two halves together so that they alternate perfectly (see figure 1). Eight perfect shuffles bring a fifty-two-card deck back to where it started. We have friends who can do this in under forty seconds, almost without glancing at the pack. To see why crooked gamblers are interested in such things, consider an ordinary pack of cards with the four aces on top. After one perfect shuffle, the aces are every second card. After two perfect shuffles, the aces are every fourth card. Thus, if four hands are dealt out, the dealer gets the four aces. It is natural to ask what can be done by combining shuffles in various ways. Is there a way we can start with four aces on top and do some combination of shuffles so that the dealer gets the aces when five hands are dealt around? While eminently practical, this last question is a math problem. We'll give the answer later when we have a few more tools.

To actually carry out a perfect shuffle is well beyond most magicians. We estimate that there are fewer than a hundred people in the world who can do eight perfect shuffles in under a minute. We're not going to try and teach this here (but see chapter 11). There *is*

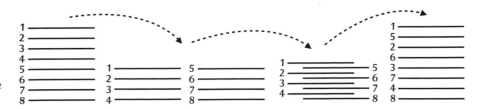

Figure 1. A perfect shuffle of an eight-card deck

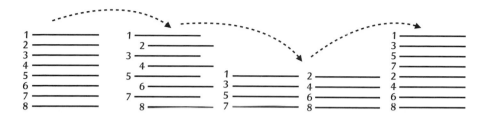

Figure 2. Reverse perfect shuffle of an eight-card deck

an easy-to-perform alternative that is useful for performing good magic tricks and has all the mathematical juice of perfect shuffles. A *reverse perfect shuffle* is shown in figure 2. When actually carried out with cards, the procedure looks like what is shown in figure 3. To try things out, take eight cards with values 1 to 8 (suits do not matter) from top to bottom. Carry out three reverse perfect shuffles, making sure to leave the original top card back on top each time. You will find the cards back in order 1 to 8. The mathematics of these reverse perfect shuffles turns out to be exactly the same as the mathematics of perfect shuffles.

Perfect shuffles are neat shuffles indeed. There are other neat shuffles that appear in card magic. Dealing cards into piles and picking up left-to-right (e.g., fifteen cards into three piles of five) is certainly neat. Dealing the top card down on the table, putting the next card under the packet, dealing the next down, the next under, and so on, is the down–and–under (or Australian) shuffle. Dealing cards from one hand to the other, putting successive cards alternately above and below previously dealt cards is called the Monge shuffle (see figure 4).

In the rest of this chapter we first show how all of these neat shuffles can be used for a solid magic trick. Following this we take a closer look at perfect shuffles. The Monge and milk shuffles are treated next. Finally, the down-and-under shuffles are discussed. Throughout, we illustrate the theory with card tricks. By the end, the reader will have a graduate course in the basic shuffles used in card magic. We will also see that the different shuffles are all part of one picture, showing the power of mathematics.

A MIND-READING COMPUTER

Make up a small deck of twelve cards as follows: Use the ace through six of spades and the ace through six of clubs. Arrange these as if

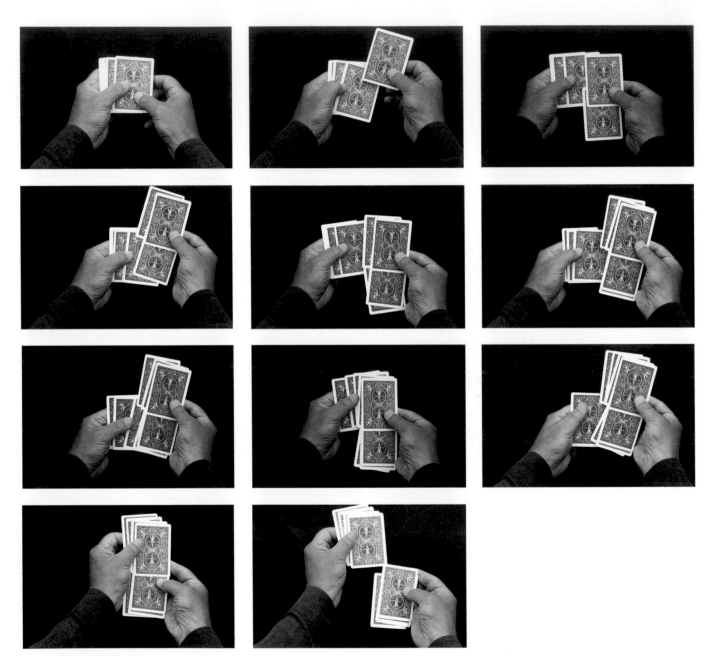

Figure 3. Doing a reverse perfect shuffle

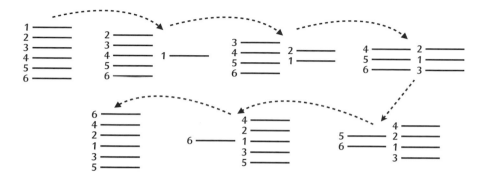

Figure 4. Monge shuffle of a six-card deck

thcy were a new deck: The ace of spades on top followed by the two of spades, and so on, ending with the six of spades. Then the clubs, in reverse order—the six of clubs, followed by thc five of clubs, and so on, with the ace of clubs on the bottom. This puts the cards into a mirror-like arrangement of a new deck. This trick can be performed as you read—go get a deck of cards and arrange them! We authors will perform for you.

The packet is going to act as a computer. It needs some input from you. Think of a small number, say, two, three, or four; deal the packet of twelve cards into that number of piles, face-down on the table, from right to left. Reassemble the piles into a single packet by putting one on the next (left to right, or right to left). Repeat once more—think of a new small number, deal that many piles, and reassemble.

The next step of mixing involves a reverse perfect shuffle as explained above. Hold the packet face-down as if you were about to deal them in a card game. Spread through the packet, pushing out alternate cards as in figure 3.

When you are through, remove the forward cards as a group and place them on top. This method of mixing is the inverse of a perfect shuffle. You have removed every other card and placed this half of the deck on top.

The packet of cards is tracking your input. Decide if you want to shuffle again, and if so, how (as another reverse perfect shuffle or again by dealing into two, three, or four piles and reassembling). Carry this out. When you are done shuffling, let the computer know by performing what we called a Monge shuffle. To do this, hold the packet in dealing position. Deal the cards from one hand to the other as follows: Push the top card off and take it in the opposite hand. Push the next card off on top of the first card. Push the next card off and put it *underneath* the first two. The next card is placed on top, the next underneath, and so on until you are done.

To review, the cards have been mixed in a way that couldn't have been predicted. The next phase involves you choosing a card and having the computer find it.

First, your choice. Hold the cards face-down in dealing position. Start moving cards one at a time from the top of the packet to the bottom. Stop whenever you want. The current top card is your choice. Deal it, face-down, off to one side. The computer will now find out which card you picked by a process of elimination.

This final elimination phase uses the down-and-under shuffle. This eliminates cards one at a time until only one remains. Proceed as follows: Hold the eleven-card packet face-down in dealing position. Deal the top card down onto the table (in a place apart from your selection). Put the next card under the packet (from top to bottom). Deal the next card onto the table, put the next under the packet and so on. This eliminates cards one at a time until one card is left in your hand.

You have selected a card freely from a mixed deck and the cards have determined a card from those remaining. The two cards should be a match in value—like the two of spades and the two of clubs. Take a look!

HOW IT WORKS. In rough outline, the trick works as follows. The small deck was initially set so that matching pairs are symmetrically arranged about the center. Thus, the top and bottom cards match, the cards second from top and second from bottom match, and so on. The cards at the center (positions six and seven) form a matching pair. This kind of symmetric arrangement is called "stay-stack" by magicians.

The first idea is this—the various methods of shuffling preserve the central symmetry. To understand, reset the cards in the initial new deck order, turn them face-up, and deal into a small number of piles. Pick up the piles in order. The symmetry is preserved. This is also true for the reverse perfect shuffle. Some further ways of shuffling that preserve symmetry are described below.

When the shuffling is finished, unbeknownst to the audience, the cards are still in stay-stack. The next shuffle, the Monge shuffle, has the effect of transforming the cards so that matching pairs are six apart. Now the deck has a different kind of symmetry. This arrangement can be cut at random without changing the fact that matching pairs are six apart.

Up to now, the deck size (twelve cards) has not been crucial. All the properties used generalize to any size deck (of even size) set initially in stay-stack order. The final down-and-under phase works as follows. Placing cards from top to bottom repeatedly is just the same as making a random cut. When the top card is placed aside as the selection, its matching mate is in the center of the remaining eleven cards. When eleven cards are processed by the down-and-under elimination deal, the central card winds up last. This final phase also works with decks of sizes 4, 12, 44, 172, . . . where the kth term of this list is obtained by adding 2^{k-1} to the previous term, starting at four. Twelve cards were used as the smallest number that allows for a reasonable trick.

MAGICAL DETAILS

1. To begin with, the cards don't have to be the ace through six. It is better to take six matching pairs at random, e.g., two red kings, two black aces, etc. These can be casually removed at the start as you explain about "programming the cards as a computer."

2. To explain the trick to the reader, we had to make the reader do all the work. It seems better to have the performer handle the cards in the initial phases of mixing and the Monge shuffle. This makes things go rapidly. Then, the packet can be handed to the spectator for the selection after a visible demonstration of how cards should be moved from top to bottom. We turn away during the selection, and then turn back, having the spectator perform the final down-and-under phase. The ending needs to be explained and built up a bit, perhaps as follows: "You have freely chosen a card from a shuffled pack. Now the pack has retaliated. If we are lucky, the pack has chosen the closest possible card to your selection. Every card has a mate—the king of hearts and the king of diamonds, the two black tens, and so on. Please turn your selection over—ah, the nine of hearts. If we are lucky, the pack should produce the matching red nine—would you turn it over please?"

3. As explained initially, the cards had to be dealt into two, three, or four piles each time. In fact, they can be dealt into any number of piles, two through twelve, and picked up to preserve central symmetry. This is a new insight, explained for the first time in this book. To

explain, say you are asked to deal into five piles. Do so, say, from right to left. Then pick up the piles in order, from left to right, with the *single exception* that the last pile is put underneath the first four.

For six cards, no special pickup is required (either left to right or right to left works). For 7, 8, 9, 10, or 11, a pickup sequence that preserves stay-stack is possible. The details are left to the reader. With a bit of practice, one can perform these pickups rapidly to look like a further randomization. Instead of asking for a number, you can ask for the spectator's name and deal one pile per letter. Thus, "Helga" would lead to five piles dealt. This kind of personalization takes the mathematical curse off the dealing.

4. An effective ploy that we have used from time to time allows the trick to be immediately repeated. At the finish of the trick, the two matching cards are face-up and the pile of ten "discarded cards" face-down. Casually pick up the ten-card packet, remove the bottommost three cards, and place them on top, reversing their order. Then pick up the two matching cards, putting one on top and the other on the bottom. You are set to repeat, starting from the beginning.

5. The presentation has been given with playing cards. The trick can be presented in many other forms. One way adopts a suggestion by Alex Elmsley. It uses twelve blank cards (like index cards) inscribed with the names of famous lovers in history. Thus, Anthony and Cleopatra, Romeo and Juliet, Liz Taylor and Richard Burton, etc. Each name is written on a separate card. At the start, the pairs are placed together. To set the trick, after showing the pairs, execute a Monge shuffle (over, under, over, etc.) "to separate the pairs." Then begin the trick as explained. One performer made up a packet of a dozen photographs and performed with these.

A different line of presentation can be based on opposites attracting. Phrases like heaven and hell, devil and angel, etc., can be inscribed on the cards.

6. For varied presentation, it may be natural to work with a different number of cards. The place that twelve is used above is in the final elimination, to wind up with the center card. There are many variations of down and under that may accommodate. For example, with a sixteen-card packet, one can begin the elimination by dealing two hands, picking up the dealer's hand and dealing it into two, and so on. This results, finally, in a single card that will be the card originally eighth from the top.

7. Dai Vernon (1894–1992), perhaps the greatest sleight-of-hand performer of the twentieth century, liked this trick and thought of a great ending. This lifts the trick out of the impromptu class but also elevates it to a very strong piece of magic. Vernon's idea shows that the final matching pair has been preordained. Since the final pair can be any of six, this necessitates having six different endings. These are best left to the individual performer. What we have in mind is something like this: After the matching pair is revealed, the performer holds a sealed envelope that contains a prediction which clearly states that the two black nines will be chosen. Of course, it may be that the prediction comes from another pocket, or the performer's wallet, etc. If the predictions can be put into full view at all times, so much the better, e.g., "I had a premonition this might happen—would you get up and look at the bottom of your chair. I taped an envelope there before the start of the show. . . ."

8. In correspondence, the great Chicago card man Ed Marlo (1913–1991) suggested a variation. This begins with the spectator cutting a twelve-card packet, then moving cards from top to bottom and stopping at will. The top card (the selection) is put aside. The remaining cards are treated by the down-and-under deal as above, resulting in one card remaining. This card is put onto the original selection, leaving ten cards. The performer now "does as the spectator did," moving cards from top to bottom, stopping after five cards have been moved. The next card is put aside. Finally, the nine-card packet is treated by the elimination procedure, starting by placing the top card *under* the packet, discarding the next, and so on. The one card remaining is placed aside. The four selected cards are now turned over—they prove to be four-of-a-kind.

The trick works as stated if the initial twelve cards are three sets of four-of-a-kind, say four aces, four twos, and four queens. These are arranged as

$$A \quad Q \quad 2 \quad A \quad Q \quad 2 \quad A \quad Q \quad 2 \quad A \quad Q \quad 2.$$

Aside from its double climax, this differs in that it eliminates the preliminary shuffling. This may be regarded as good by some who have the ability to substitute some sleight of hand. It is possible to do certain shuffles that retain the arrangement described above. For example, the packet can be reverse perfect shuffled in groups of three, or it can be dealt into five piles, which are assembled in order 1 3 5 2 4. This last, a subtlety of the underground California card man Bob Page, is capable of broad generalization.

A BIT OF HISTORY

The trick described above assumed its shape over a twenty-year period. It began with a problem posed by magician Bob Veeser to Ed Marlo. Marlo worked out a way to do the basic effect using sleight of hand and published a version in the American magic magazine *Tops* in November 1967. On reading this, we figured out the non–sleight-of-hand version described above and communicated it to Marlo in that same month. Magicians have an active network. Before the month was out, Marlo sent back the four-of-a-kind variation described above. From that day to this, we have enjoyed performing the trick and developing some of the variants detailed above.

Magic is a world that revolves around secrets. We showed this secret to a nameless cad who communicated it to a prolific writer, Karl Fulves. Fulves used Marlo's version of the trick (without permission) in a book for the public called *The Magic Book*. Fulves has infuriated some magicians by publishing cherished secrets without permission or credit. Of course, this also means his books have some good magic in them.

A LOOK INSIDE PERFECT SHUFFLES

Our own introduction to the mathematics of perfect shuffles came through cutting school and hanging around New York magic stores in 1954. We met Alex Elmsley, a brilliant young inventor of magic tricks who was visiting America from England. He explained that there are two kinds of perfect shuffles—the "out-shuffle" and the "in-shuffle." The out-shuffle leaves the original top card back on top. The in-shuffle begins by splitting the deck into two equal piles and shuffling perfectly, leaving the original top card second from top. The two kinds of shuffles are similar, but you will find it takes six in-shuffles to recycle an eight-card deck (versus three out-shuffles). It takes fifty-two in-shuffles to recycle a fifty-two-card deck. We are embarrassed to report that we discovered this by actually shuffling the cards. Non-sleight of handers can easily follow along by noting that there are also two types of reverse perfect shuffles (see figure 5) determined by the original top card either winding up on top or not.

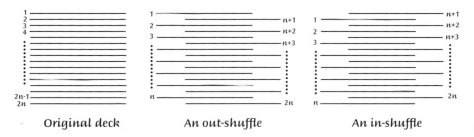

Original deck   An out-shuffle   An in-shuffle

Figure 5. Out-shuffles and in-shuffles of a deck of $2n$ cards

Alex Elmsley had discovered that by *combining* ins and outs, you can do amazing things. For example, it is useful for a magician to be able to bring the original top card to a given position (say, thirteen) by a sequence of shuffles. Can this always be done? If so, how? Here is his method. Subtract one from the final position (so $13-1=12$). Express this last number in binary arithmetic, using zeros and ones (so $12=1100$). Interpret the ones as in-shuffles and the zeros as out-shuffles, left to right. This sequence of shuffles does the job (in, in, out, out brings the original top card to position thirteen). This was the thirteen-year-old member of your author team's introduction to binary arithmetic. It seemed just as magical as the tricks themselves.

It is natural to ask just what can (and can't) be done with the two types of perfect shuffles. Can *any* arrangement be realized? Can we shuffle in a way that starts with the four aces on top and winds up with the aces at positions 5, 10, 20, 25 (so they come to the dealer's hand if five poker hands are dealt)?

We solved this problem in joint work with William Kantor, a University of Oregon math professor. It turned out to be months of hard work for the three of us. To explain the results, we have to use the idea of a centrally symmetric arrangement (which we also described in the previous section). Consider the following arrangement of an eight-card deck: A B C D D C B A. The two As (top and bottom cards) are symmetric about the center. Similarly, the two Bs, the two Cs, and the two Ds. With a fifty-two-card deck, the top and bottom, the second from top and second from bottom, and the middle pair (26 and 27) are centrally symmetric pairs. A Philadelphia policemen, J. Russell Duck (who published as Rusduck), discovered the following invariant of both in- and out-shuffles. He is the one who coined the term "stay-stack." (Incidentally, he also started the first magic journal, *Cardiste*, devoted exclusively to card magic.) Following either kind of shuffle, centrally symmetrical pairs stay centrally symmetric. Thus, consider the original top and bottom cards. After *any* sequence of in- and out-shuffles, they must occupy centrally symmetric positions. The same holds for any starting symmetrical pair.

One reason for undertaking our study of just what is possible with in- and out-shuffles is to see if there are other useful invariants. What we have so far is that only rearrangements preserving central symmetry are possible. It follows that there is no sequence of in-shuffles and out-shuffles that results in the top two cards being transposed and the rest of the deck remaining in order. (Actually, there is one exception to this statement. Can the reader see what it is?)

Our study showed that, in a rough sense, magicians had found all the hidden symmetries. Decks of size a power of two (for example, eight or thirty-two) have extra symmetries (a lot of them!), but aside from these and a few exceptions for decks of size twenty-four or smaller, the only pattern preserved by both in- and out-shuffles is stay-stack *and* a simple parity condition that cuts things down by a factor of two (or four, when both the full permutation and the permutation of pairs contribute). For now, let us just say that for decks of size 52, 60, 68, . . . , there is no parity consideration. Rusduck's stay-stack is all that is preserved. *Any* arrangement consistent with stay-stack is achievable by a sequence of in- and out-shuffles. In particular, consider again the four aces on top of a fifty-two-card deck. Earlier, we asked if there were some sequence of shuffles that stacks them for a five-handed poker game. That is, puts the aces in positions 5, 10, 15, and 20. Since there are *many* ways of completing these positions consistent with stay-stack, there is indeed some way to get there using in- and out-shuffles. One can even make the aces come out in a known order. We hasten to add that we have no idea what such a sequence is, or even what the length of the shortest possible sequence achieving this is.

Those wishing to study perfect shuffles in more depth are encouraged to get a copy of S. Brent Morris' book *Magic Tricks, Card Shuffling, and Dynamic Computer Memories*.[2] This treats the history, tries to teach perfect shuffles, and has further tricks. It also contains further applications to computer hardware and software design.

Just to show that the field is still active, we record the latest breakthrough. Consider the following *inverse* problem: What sequence of shuffles is required to bring a given card to the top? This problem, posed by Alex Elmsley in 1958, appeared to have no neat solution. It can always be done, but the actual sequence depends on the size of the deck in a seemingly unknowable way. Indeed, magicians can purchase a book listing the required patterns for various deck sizes. One easy case for the inverse problem occurs for decks of size a power of two (e.g., eight-card decks or thirty-two-card decks). Paul Swinford, a Cincinnati card man, discovered that for such deck sizes, the pattern of shuffles that brings the top card to a given position *also* brings the card in that position to the top.

However, computer scientists S. Ramnath and D. Scully recently discovered a solution.[3] Here is our version of how they do it. The following may appear to be cryptic. More details are in our paper "The Solutions to Elmsley's Problem."[4] Suppose we start with a deck of $2n$ cards, and we label their initial positions $0, 1, \ldots, 2n - 1$ (i.e., the top card is in position 0). To determine the shuffle sequence that will bring the card in position p to the top, we do the following. First, let r be the integer that satisfies $2^{r-1} < 2n \leq 2^r$. Note that we can assume that $0 < p < 2n - 1$, since if $p = 0$ then we don't have to do anything, and if $p = 2n - 1$ then r consecutive in-shuffles does the job. Next, let t be the largest integer not exceeding $2^r(p + 1)/2n$, and write $t = t_1 t_2 \ldots t_r$ in its base two expansion. For the next step, let $= s_1 s_2 \ldots s_r$ be the last r digits of the base two expansion of $2nt$. Finally, form the binary sequence $u_1 u_2 \ldots u_r$ by defining $u_i = s_i + t_i$ where the addition is done modulo two. Then the desired shuffle sequence can be read off from left to right from the u_i by interpreting 0 as an out-shuffle and 1 as an in-shuffle.

As an example, suppose our deck has $2n = 52$ cards, and we would like to bring the card in position 37 to the top. In this case, $r = 6$ since $2^5 < 52 \leq 2^6 = 64$. Thus, since $64 \times 38/52 = 46 + 10/13$, then $t = 46 = 101110$ base two. Also, since $52 \times 46 = 100101011000$ in binary, then the last six digits of this are 011000. Thus, the u_i sequence is the modulo two sum 110110, and so the desired shuffle sequence is I I O I I O (i.e., in, in, out, in, in, out). Note that the last shuffle is superfluous. We admit that this computation might be difficult to perform in real time, so to speak. Perhaps the reader can find a simpler way to explain it or, better yet, an even better way of producing the desired shuffle sequence.

A LOOK INSIDE MONGE AND MILK SHUFFLES

The Monge (or over/under) shuffle is carried out by successively putting cards over and under (see figure 6). Thus, the top card is taken into the other hand, the next is placed above, the third below these two, and so on. For example, eight cards in order 1, 2, 3, 4, 5, 6, 7, 8 from top down wind up in order 8, 6, 4, 2, 1, 3, 5, 7 after one Monge shuffle. There is also a "down" Monge shuffle. Here, the top card is taken into the other hand, the second card is placed under this, the third card above, and so on. Eight cards wind up in the order 7, 5, 3, 1, 2, 4, 6, 8 after a down Monge shuffle. The milk (also known as the Klondike) shuffle involves successively sliding off the current top and bottom cards of the packet, dropping these pairs onto a single pile on the table as they are "milked" off. Eight cards milk shuffled end up in the order 4, 5, 3, 6, 2, 7, 1, 8.

The two shuffles are inverses of one another. Thus, a milk followed by a down Monge (or vice versa) leaves the original order unaltered. This means that many properties of milk and Monge shuffles coincide, such as their cycles and orders.

Monge shuffles are frequently used to reorder a pack. For example, if cards are initially arranged in matching pairs (e.g., two red queens on top, two black sevens next, etc.), after one Monge shuffle they are in a reflection (or centrally symmetric) arrangement Q, 7, . . . , 7, Q. After a second Monge shuffle, they are in a parallel arrangement, with the matched pairs exactly half a deck apart. Our computer mind-reading trick gives a magical example of the utility of switching from reflection to parallel.

Gaspard Monge (1746–1818) was an eighteenth-century geometer, mainly remembered today for Monge cones, which are geometrical objects associated with partial differential equations. He worked out the basic mathematical details of these shuffles in 1773.

The milk, or Klondike, shuffle can also be used for tricks. An early example of the combination of Monge and milk shuffles for cheating at cards appeared in the anonymously authored *Whole Art and Mystery of Modern Gaming* in 1726.[5] The author begins with a deck arranged ace through king four times over, with suits alternated. The top thirteen cards are over/under shuffled and placed on the table. Follow this by over/under shuffling the next thirteen cards, and so on. The four thirteen-card packets can be assembled in any order. Following

Figure 6. Doing a Monge shuffle

this, milk shuffle the entire deck. Now the cards can be repeatedly cut. The resulting arrangement guarantees success at the game of faro. In this game, as popular in its day (circa 1750–1850) as casino blackjack is today, the player chooses some value and pairs of cards are dealt off; the first card of the pair wins (if the value is matched), while the second card of the pair loses (if the value is matched). If two equal values come up (a split), the house takes half of all bets on this value. Players can bet on any value to win (or lose). With the careful shuffling described above, each value alternates between winning and losing. Because of the cut, players cannot predict the first appearance of a value but, following this, if the first appearance wins, the next will lose, the next will win, and the last will lose. Since bets can be made at any time, a knowledgeable player in cahoots with the dealer can clean up.

Formulas for the order (that is, how many times the deck must be shuffled to recycle) and position of the top card after a milk or Monge shuffle are easy to work out. However, they also follow from what we know about perfect shuffles as explained below.

A LOOK INSIDE DOWN-AND-UNDER SHUFFLES

The down-and-under shuffle is a familiar elimination procedure with ancient precursors going back to the Roman historian Flavius Josephus. It affects all of us when we play "(S)he loves me, (s)he loves me not" with the petals of a daisy. In its simplest form, a packet of n cards has its top card placed down on the table, its next placed under on the bottom of the packet, the next down, the next under, and so on, until just one card remains (see figure 7).

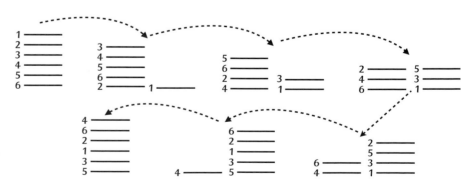

Figure 7. A down-and-under shuffle with a six-card deck

The question, "What card winds up last?" has an easy answer: Subtract from n the highest power of two that is less than n, and double this number. Thus, if $n = 12$, the highest power of two less than 12 is 8, so $12 - 8 = 4$ and $2 \times 4 = 8$. So in this case, this last remaining card is the card that was originally in position 8.

There are many fine tricks that place a chosen card into the right position. Here is one of ours. Have the spectator shuffle a packet of 2^n cards. Deal part of the packet alternately into two piles of equal size. The spectator may choose either of the tabled piles or the packet remaining in his or her hand. If a tabled pile is chosen, the spectator looks at (and remembers) the bottom card in the packet remaining in hand and drops this packet on top of the chosen tabled packet, discarding the remaining packet. A down-and-under deal on the combined packet will reveal the chosen card as last (see figure 8). If the nontabled packet is chosen, the spectator looks at the bottom card and drops this packet on *either* of the tabled packets, and then does a down-and-under deal. Put this bland way, the trick is poor. Can the reader dress it up to make something performable?

There is an extensive mathematical development and more careful history of down-and-under shuffles in Herstein and Kaplansky's fine book, *Matters Mathematical*.[6] The connection to perfect shuffles is developed in the following section.

ALL THE SHUFFLES ARE RELATED

So far, we have introduced in- and out-shuffles, reverse in- and out-shuffles, the Monge shuffle, the milk shuffle, and the down-and-under shuffle. Of course, there is a natural pairing between the out- and reverse out-shuffles. If you do a perfect out-shuffle and follow it with a perfect reverse out-shuffle, the cards return to their original order. The two shuffles serve as each other's inverse. This means that the properties of repeated out-shuffles can be deduced directly from the properties of repeated reverse out-shuffles. Indeed, if a deck of cards recycles after K out-shuffles, then the actual arrangement after any number M of out-shuffles is exactly the same as the arrangement after $K - M$ reverse out-shuffles. For fifty-two cards, eight out-shuffles recycle the deck, so K is eight, and therefore two out-shuffles leave the cards in exactly the same arrangement as six reverse out-shuffles.

Figure 8. Doing a down-and-under shuffle

The relation between in- and out-shuffles is simple too: After an out-shuffle, the top and bottom cards stay at the top and bottom (of course). The rest of the deck's cards move the same as an in-shuffle of a deck with two fewer cards. Because of this, properties of repeated in-shuffles are completely determined by properties of repeated out-shuffles. For example, eight in-shuffles recycle a deck of fifty cards.

The relation between perfect shuffles and Monge shuffles (over, under, over, under, . . .) and their inverses, milk shuffles, explained above, is more subtle but still easy. The alert reader may have noticed there are two types of Monge shuffles—after we deal off the top card, the next card can be put *over* (and then continuing under, over, etc.). Call this an "up" shuffle. Alternatively, after we deal off the top card, the next card can be put *under* (then continuing over, under, over, under, etc.). Call this a "down" shuffle. Can the reader give a simple description of an *inverse* down shuffle?

On the surface, Monge and milk shuffles seem quite different from in- and out-shuffles. Here is the connection (pointed out to us by the Princeton mathematician John Conway): In an out-shuffle the symmetrically located pairs (top and bottom, second from top and second from bottom, etc.) stay paired. However, these pairs move around relative to each other. For example, twelve cards arranged A B C D E F F E D C B A are out-shuffled to A F B E C D D C E B F A. Since the top half determines the bottom half, we may study how the pairs permute by just following the top half. This is:

$$A \; B \; C \; D \; E \; F \to A \; F \; B \; E \; C \; D.$$

This is just a milk shuffle with the cards face-up to start (so card F starts on top). All of this means that properties of Monge and milk shuffles are equivalent to properties of out-shuffles. It means that patterns and calculations found over two hundred years of study of Monge shuffles apply to out-shuffles and that our theorem, determining all possibilities when in- and out-shuffles are combined, determines all possible arrangements when the two types of Monge shuffles are combined.

The connection between perfect shuffles and the down-and-under shuffles lies further beneath the surface. It was discovered by Paul Lévy, the great French probabilist. He was confined to bed as a child for many weeks and found solace in fooling around with simple card tricks. Years later, when confined to bed as an older man, he again began fooling around with simple shuffles. He discovered that the structure of a down-and-under shuffle of a deck of size n is the same as the structure of a milk shuffle, provided that $2n - 1$ divides a number of the form $2^t + 1$. The definition of structure (technically, *cycle* structure) would take us too far afield, but "same structure" means that, after relabeling the cards, the two arrangements are the same. In particular, the number of repetitions required for the deck to return to its original order is the same. For example, a milk shuffle of five cards and a down-and-under shuffle of five cards each fix two positions and cycle the other three. In particular, both recycle after three repetitions. Here, $2n - 1 = 9$, which divides $2^3 + 1$. We do not know if there is a connection between down-and-under shuffles and perfect shuffles for decks of more general size.[7]

To conclude, we observe that there are many variations possible. As an example, an early conjuring book, John Gale's *Cabinet of Knowledge,*[8] presents a bevy of tricks and gambling demonstrations based on shuffling cards by pushing off two cards in a group, then the next three (in a group) on top, the next two underneath, and so on, alternating by twos and threes. Gale remarks that an n-card deck recycles after $n - f$ shuffles with f being the number of fixed points of the original shuffle.[9] As far as we know, Gale's shuffles have not been studied further. With obvious variations, they suggest new territory to be explored.

Neat shuffles are a real marriage of magic and mathematics. Discoveries have come from both sides. One has to be careful, as the tricks are close to the boring, dealing-into-piles tricks that everyone dreads. With work and luck we are sure there is more gold to be discovered. We hope the reader finds some good tricks and enjoys the mathematical mortar as well.

Chapter 7

THE OLDEST MATHEMATICAL ENTERTAINMENT?

A thirteen-year-old boy slowly opens the door to the world's largest magic shop. It's two in the afternoon and the boy has cut school, making the trip on New York's grimy subways. The shop is Louis Tannen's Magic Emporium at Forty-second Street and Sixth Avenue in New York's Times Square. Not the usual street-level shop with doggie doo and plastic vomit in the window, Tannen's is on the twelfth floor of the Wurlitzer Building. You have to know about it to find your way in. It's a slow day, but some of the regulars look up and smile. There's Manny Kraut, a huge man whose fat hands somehow make the most beautiful, delicate trick cards. There's Harry Drillenger, a retiree who used to play the musical saw in vaudeville. He loves flourishes with cards and has taught a few to the boy. One time, when cards had accidentally been dropped on the floor, Harry picked them up, one by one, carefully wiping them clean with a silk scarf he always wore. "Cards don't like to be dropped," he admonished.

The owner of the shop came out to see who had walked in. He was Lou Tannen himself, a red-headed pixie whose eyes twinkled. When he demonstrated tricks, he whistled silently. "Hey, I've got something for you," Tannen said to the boy. Happy to be recognized, the boy shyly approached the counter. "You're our best young sleight of hander," Tannen said. "We're giving you a prize." He pulled out a new five-dollar bill and said, "You can buy whatever you like." Five dollars was a lot of money for the boy. The subway cost fifteen cents each way then, and some days the boy would run low on money and walk back from Forty-second Street to Two-hundredth Street.

The boy looked around the shop with very wide eyes. He saw crystal caskets, bright yellow tubes imprinted with stenciled Chinese characters, he saw really thousands of things. Behind the counter were the books. Magic books that carefully told the secrets. An amazing set of books had just been remaindered. John Northern Hilliard's *Greater Magic* was by far the biggest and deepest tome ever. A publisher had put it out in five volumes, hoping to have a best seller. It hadn't worked out and Tannen's had the set of five for two dollars a volume. After a long time, the boy asked, "Could I buy *Greater Magic*?" Tannen adopted his shrewd look. He knew the boy had no extra money and offered, "Well, I'll knock it down to eight for you." The boy's face fell; he didn't have the extra three dollars. "Oh, go ahead and take them," Tannen said gruffly, handing the boy the treasures and pocketing the five. Harry and Manny smiled.

What treasures they were. Hilliard had been a newspaper man and then the advance man for America's largest magic show, Howard Thurston's Magical Extravaganza, which travelled from city to city, week in and week out. Hilliard would arrive a week in advance, arrange publicity with the newspapers, set up interviews, and talk the show up with the local magicians. Despite it being his day job, he loved magic and the crazy magicians. He collected the best and the brightest, the most secret of secrets, and he wrote them up over a thirty-year period. This became *Greater Magic*.

Hilliard had a grand style and could invoke the mysteries of the ages without seeming too corny. One chapter was called "Old Wine in New Bottles." It begins with the quotation, "When a new trick comes out, I do an old one." It continues, "We that are old—the phrase is Richard Steele's, not mine—have seen so many magical effects arise and then slowly sink down, fade out and pass into obscure interment, the limbo of forgotten things, that we are not going to pray your patience and forbearance for this chapter, which is, in a manner of speaking, an adventure in antiquity. The 'recreations' and 'amusements' in the quaint terminology of the old chroniclers—in the pages that follow are among the oldest effects in legerdemain."

Hilliard goes on to sketch out his version of the history of magic and then to present some of the crown jewels of magic. Hilliard's book is still in print. It has been gloriously republished with some missing chapters and background. It you want to intrigue a youngster of any age, buy a copy of *Greater Magic.*

One of the jewels that Hilliard described in his "Old Wine" chapter—"glazed with the patina of age"—he called the Miracle Divination. It was one of the boy's first mathematical tricks. It still makes for terrific magic. Here is the way we do it today.

THE MIRACLE DIVINATION

To perform this trick, you will need twenty-four pennies and a nickel, dime, and quarter. Sixty-four cents in all. You can keep them together in a small purse and then just dump them into a pile on the table. Casually pick up six of the pennies. If it is your style, proceed as follows: "You know, it takes money to make money. Who would like to help us think about making money?" Three spectators are brought up. "I'm going to start each of you off with some capital." Hand the first spectator a penny, the second spectator two pennies, and the third spectator three pennies. This should be done casually, without comment—the spectators shouldn't be aware that they have different amounts. Push the nickel, dime, and quarter forward on the table. Address the three spectators (call them John, Mary, and Susan) as follows: "First, a lesson in leveraged mortgages. I'll turn my back and, while I promise not to peek, you should all keep your eyes on me. John, the three coins on the table represent three properties you can buy. I'd like you to pick up one of them. Mary, there are two options left. Please make a free choice, high or low. Susan, you get what's left—what can I say, it's better than nothing." With your back still turned, continue the patter as follows: "There will be points on our loans. Whoever picked up the nickel, that's our smallest property. Please take the same number of pennies from the pile as the number I gave you originally. Now, one of you has the dime. That's a larger property. Please take twice as many pennies as I gave you. Finally, whoever took the quarter, please take four times as many pennies as I gave you, as a reward for having grand visions."

The performer keeps talking, "Now, I'm bankrolling this operation, and one of the keys to being a successful banker is to know your customer. John, you made the first choice. Let's see, are you a risk taker?" While talking, turn to face the audience, casually sweep the remaining pennies into your hand, and dump them in a pocket. Of course, you will have secretly counted them and (as explained below) this tells you who took what. "John, were you watching me? Did I peek? Let's see, I'll have to use psychology—you couldn't make up your mind between small and large—you took the dime. Mary, you played it big—you picked up the quarter, is that right? Susan, you were left with the nickel. It's only small change—better luck next time."

THE EFFECT

COUNTING THE WAYS. The performer initially puts twenty-four pennies in a pile on the table. From these, six are removed: One is given to the first spectator, two are given to the second spectator, and three are given to the third spectator. No particular mention is made of these numbers. Proceeding as explained, at the end the performer finally turns around and counts how many pennies are left. The number will be 1, 2, 3, 5, 6, or 7, and this result uniquely codes all the spectators' choices. If the three original objects are coded as N, D, and Q, the chart in table 1 tells each spectator's choice. For example, the first row codes the case in which Spectator 1 takes the nickel, Spectator 2 takes the dime, and Spectator 3 takes the quarter. Following the instructions, Spectator 1 removes one more penny, Spectator 2 removes four more (twice as many as originally given), and Spectator 3 removes twelve more (four times as many). Thus, $1 + 4 + 12 = 17$ are removed from the eighteen left on the table, so one penny is left. As we see from the table, this codes $\frac{1}{N}\frac{2}{D}\frac{3}{Q}$. The other rows can be similarly checked. Practical performance details are given below, but first we give the history of this first trick!

Table 1. Coding the remainder for the spectators' choices

Number left	Spectators' choices		
	1	2	3
1	N	D	Q
2	D	N	Q
3	N	Q	D
5	D	Q	N
6	Q	N	D
7	Q	D	N

SOME HISTORY. The history of magic contains its own mysteries. One remarkable coincidence that we do know: The first two serious printed magic books came out in the same year—1584. The books appeared in France and England, respectively, and have completely different contents. The first, J. Prevost's *La Premiere Partie des Subtiles et Plaisantes Invention*, published in Lyon, tries to teach magic in some detail—effects, methods, and what to say, all described in a leisurely fashion. Prevost describes two versions of the Miracle Divination.

The other 1584 magic book, Reginald Scot's *Discoverie of Witchcraft*, was mainly written to protest the growing mistreatment of the old and frail as witches. Along the way, he includes a chapter describing magic tricks. Scot's book contains a different set of sleight-of-hand tricks and no mathematical tricks. The French and English traditions evolved quite differently. In England, later magic books copied Scot's book for one hundred fifty years with but small variations and additions. In France, there was much greater variety and a definite attempt to improve on the old, to look back and unify past achievements.

One clear line of development can be seen by following the progress of the *three-object divination*. The second French magic book appears in 1612. This is Gaspard Bachet's *Problemes Plaisants et Delectables Qui se sont par les Nombres* (also published in Lyon). Bachet describes the three-object divination and then goes on to prove what may be the first theorem of mathematical magic: He shows that a straightforward generalization to four or more objects will not work. He gives a four-object version along with a fantastic early example of what we called generalized de Bruijn sequences in chapter 4. We pause to describe Bachet's results.

The four-object trick involves four spectators initially given 1, 2, 3, and 4 counters, respectively. First, they each choose one of four objects while the performer's back is turned. Then, whoever took object A takes as many more counters as originally given. Whoever took B takes four times as many as originally given. Whoever took C takes sixteen times as many as originally given. (D gets nothing more.)

The number of counters remaining in the pile on the table again uniquely codes the objects chosen. In addition to giving a table similar to table 1, Bachet presents a circular arrangement (see figure 1). An outer circle shows the twenty-four numbers that can arise as remainders. The inner circle codes the objects chosen in a clever way: Under the remainder actually occurring is a 1, 2, 3, or 4, signifying what the first person chose. The next number, going down around the circle, shows what the second spectator chose and the third number shows what the third spectator chose. This determines what the fourth spectator chose. This arrangement is the earliest known occurrence of a generalized de Bruijn sequence.

Bachet's second contribution gives counterexamples to the claims of earlier writers who believed that a straightforward extension of the three-object divination works out. Bachet shows by example that the obvious extension (five people given 1, 2, 3, 4, and 5 counters, respectively, and the person choosing the ith object takes 2^{i-1} times what was originally given) does not give a unique coding. The permutations: $\begin{smallmatrix} 1 & 2 & 3 & 4 & 5 \\ a & b & c & d & e \end{smallmatrix}$ and $\begin{smallmatrix} 1 & 2 & 3 & 4 & 5 \\ a & c & d & b & e \end{smallmatrix}$ give the same number—121 removed—and so the same remainder. In fact, the four-object case with this coding

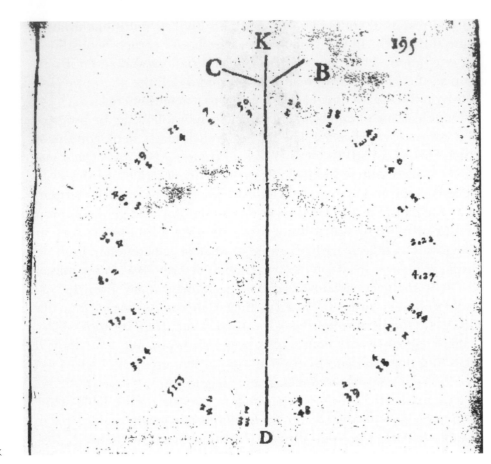

Figure 1. A page from Bachet's book

fails as well. Can the reader see why? Bachet's version is different and his four-object version works fine. Let us return to performance magic.

RON'S $1.96 TRICK

Bachet's four-object divination required a special table lookup. We next describe a performable four-object divination that can be done by "pure thought."

THE TRICK. The performer dumps a pile of change onto the table and patters as follows. "The way we think about money in this society is complex: We would all like more, yet most of us do not want to look greedy. I have turned my back and want four of you to assist me. Allen, please pick up any single coin from the pile on the table. Betty, I want you to select a coin too, but choose one with a value different from Allen's. Charlie, would you please pick up one of the coins, again of different value from what Allen and Betty chose. Finally, Diane, I hope you have been watching. I would like you to choose a coin different from what the others have chosen."

The performer continues: "As further reinforcement for your thoughts, I want you to take some more money from the table. This time it does not have to be a single coin. Diane, you chose last. What-ever you took before, now take four times that amount. You can take or give change if that will help—so if you took the dime, take forty cents more. Allen, you went first. I'm not sure what you took, but probably you were greedy—take as much more from the table as you first took.

"Betty, please take double the amount you took. Charlie, you also chose late. Please take three times the amount you first took—so if you took the nickel, take fifteen cents more."

At this stage, the performer turns back around and, without asking questions, reveals all four spectators' choices. For each, this can be either the amount originally chosen, or the total amount in hand. The presen-tation can be varied. One can analyze their personalities, as in, "Allen, you are a subtle man—you did not want to choose a quarter, too greedy, but the opposite is too obvious too—you picked up a dime." For other spectators you might patter as follows: "Charlie, I cannot picture what you did at the moment you chose, but do you have all your money in your fist? Let me take your wrist and listen— that sounds like twenty cents."

HOW IT WORKS. To begin with, the performer dumps $1.96 on the table—six pennies, six nickels, six dimes, and four quarters. The trick has been designed so that the amount left on the table codes the spectators' choices. Further, we have created this variation so that you can easily unravel the information in your head, without consulting a chart or memorizing anything.

First, figure out how much is left on the table. Call this total amount T for total. If the four spectators are called A, B, C, D (Allen is A, Betty is B, Charlie is C, and Diane is D in our example), then we will use the code $A = 4$, $B = 3$, $C = 2$, $D = 1$.

Step 1: Reduce T modulo 5. The remainder will be one of 1, 2, 3, 4, which will tell you who took the *penny*.

Step 2: Reduce T modulo 4. The remainder r taken in 1, 2, 3,4 codes who took the *dime*. Thus, remainder 4 means Spectator A took the dime, and so on.

Step 3: Divide T by 5, ignoring the remainder, and add this to $3 \times r$. The result modulo 5 tells you who took the *nickel*.

Step 4: The remaining spectator took the *quarter*.

AN EXAMPLE. Suppose the choices initially made are

A	B	C	D
nickel	*quarter*	*dime*	*penny.*

Following your instructions, A removes five cents more, B removes fifty cents more, C removes thirty cents more, and D removes four cents more. This makes an additional eighty-nine cents removed. When you turn back around you see sixty-six cents, which is what we call the total T.

Step 1: 66 modulo 5 is 1, so Spectator D took the penny.

Step 2: 66 modulo 4 is 2, so Spectator C took the dime.

Step 3: $\frac{66}{5}$ rounded down is 13. Thus, $13 + 3 \times 2$ (from step 2) equals 4 modulo 5, so Spectator A took the nickel.

Step 4: The remaining spectator, Spectator B, took the quarter.

WHY IT WORKS. First, observe that if a spectator took x coins during the second phase of the trick ($x = 1$, 2, 3, or 4), then the *code* for that spectator is $5 - x$. After the first phase, a total of $1 + 5 + 10 + 25 = 41$ cents has been removed from the table, leaving \$1.55, or 155 cents. Suppose that for the second phase, p pennies, n nickels, d dimes, and q quarters have been removed. Hence, the amount left on the table is

$$T = 155 - (p + 5n + 10d + 25q). \qquad (7.1)$$

For step 1, we reduce T modulo 5 to get $T \equiv -p \pmod 5$. Thus, $p \equiv -T \pmod 5$, and so the code for the spectator who took the *penny* is $5 - p \equiv T \pmod 5$.

For step 2, we reduce T modulo 4 to get $T \equiv 3 - (p + n + 2d + q) \pmod 4$. However, we now use the fact that exactly ten coins have been removed from the table during the second phase, so that $p + n + d + q = 10$. Plugging this in, we get $T \equiv 1 - d \pmod 4$, i.e., $d \equiv 1 - T \pmod 4$. Hence, the code for the spectator who took the *dime* is $r \equiv 5 - (1 - T) \equiv T \pmod 4$.

For step 3, we divide equation (7.1) by 5 to get

$$\frac{T}{5} = 31 - \frac{p}{5} - n - 2d - 5q. \qquad (7.2)$$

Since $p = 1, 2, 3$, or 4, when we round down $\frac{T}{5}$ to the next whole number, which we denote by $\lfloor \frac{T}{5} \rfloor$, we have

$$\left\lfloor \frac{T}{5} \right\rfloor = 30 - n - 2d - 5q.$$

Now, reducing this equation modulo 5, we get

$$\left\lfloor \frac{T}{5} \right\rfloor \equiv -n - 2d \pmod 5,$$

i.e.,

$$n \equiv -\left\lfloor \frac{T}{5} \right\rfloor - 2d \pmod 5.$$

So, the code for the spectator taking the *nickel* is $5 - (\lfloor \frac{T}{5} \rfloor - 2d) \equiv \lfloor \frac{T}{5} \rfloor + 2d \pmod 5 \equiv \lfloor \frac{T}{5} \rfloor + 3r \pmod 5$, as claimed. The remaining spectator took the *quarter*.

This version of the trick is eminently performable. It will take work to familiarize yourself with the rules and to get facility with the division steps involved. An easy way (in step 3) to calculate the remainder of T divided by 5 is just to *double* T and divide by 10. For example, $\frac{66}{5} = \frac{2 \times 66}{10} = \frac{132}{10}$, which rounds down to 13. At first this can be slow and painful, but with practice it will flow smoothly. In performance, we have found that a tiny bit of mnemonics helps.

penny	copper-policeman	(copper *penny*)
nickel	nickelodeon	(*nickel*odeon)
dime	tin can	(one thin *dime*)
quarter	stick	(*quarter* staff)

Using mnemonics, one creates vivid visual images that the mind is able to recall easily. The list above associates easy-to-picture objects with the four coins. As you go through the trick, you will deduce choices sequentially. For example, you may deduce that Charlie took the penny. There are other computations to come and you want to keep this in mind. Picture Charlie in a battle with a policeman—fisticuffs flying. It takes an instant and will stick with you. If Diane took the dime, picture her crushing the tin can. The more vivid and strange a picture you make, the easier it is to recall.

VARIATIONS. Over the years, magicians have performed hundreds of variations of the basic three-object divination. Some performances are supported by elaborate plots. One such performance involves a murder mystery, the basic objects being a knife, a gun, and a noose. This kind of presentation allows for performance in theaters for huge audiences. At the other extreme, Californian Glenn Gravatt, a detective in real life, performs a version involving a single spectator and a one-, five-, and ten-dollar bill. The spectator puts the bills into three separate pockets and, after following some simple instructions, Glenn determines the bills' locations by the number of coins left on the table.[1]

One problem that intrigues us: What are some subtle ways of finding out the total? In our version, we have to count the remainder. Spectators may notice this, giving them a clue to the way the trick works. We have thought of having coins in a bag and weighing what is left. We have thought of dumping the handful of change remaining into a glass of black ink and using the displacement of the liquid to tell the total. We are sure there is a great simple variant to be found. We hope readers will tell us if they find an elegant solution.

EARLY MAGIC

The history of the three-object divination really touched us when one of us acquired a copy of the first serious French magic book— J. Prevost's aforementioned *La Premiere Partie des Subtiles et Plaisantes Inventions*. There are fewer than ten known copies of the book extant; holding the actual volume makes its tricks come to life in a way that microfilms or reproductions can't. Maybe the author handled *our* copy; it is over four hundred and twenty-five years old as we write this.[2] In his book, Prevost gives complete performance details for the three-object divination, including the advice to draw the relevant table of permutations upon the palm of your hand. He also gives a variation using thirty counters, which has disappeared from later literature.

With the history made real, the search was on. Was Prevost's description the earliest? For a long time, we didn't know. One solid premise concerning history is this: Never accept a clear first appearance. The history of magic is not a well-studied area. Magicians have done the best they can but magicians simply aren't trained scholars. Nevertheless, a band of dedicated amateurs has begun to make progress in pushing back the date of the earliest known appearance. While Prevost's and Scot's books are the two most important early "serious magic books," a picture has emerged of extensive bits and pieces published earlier. The scholarly finding of magicians Bill Kalush and Vanni Bossi led to many other early discoveries.[3] The best history of magic is the newly issued second edition of S. W. Clarke's *The Annals of Conjuring*.[4] The appendices, particularly those authored by Hjalmar and Thierry dePaulis, identify numerous pre-1584 items.

Our breakthrough in tracing the history of the three-object divination came only recently (we had been looking for more than twenty years). A true historian, Albrecht Heeffer, wrote a paper titled *"Récréations Mathèmatiques* (1624): A Study on Its Authorship, Sources, and Influence."[5] Heeffer's immediate object of study is the first-known book to use the title *Mathematical Recreations*. This appeared anonymously in 1624 and was reprinted in many subsequent editions. Scholars have argued vehemently over its authorship. We will not enter that labyrinth here but encourage the reader to read Heeffer's wonderful overview. In the course of his work, Heeffer traces the lineage of many standard tricks. From Heeffer we found earlier appearances of the three-object divination and some later scholarship. The trick was a standard item in books teaching arithmetic.

After Heeffer opened our minds, we looked back at the standard "earliest source" for modern arithmetic. This is Fibonacci's *Liber Abaci*, published in 1202. (Fibonacci was also known as Leonardo Pisano.) This is finally available in a modern English edition.[6] In the eighth part of the twelfth chapter of this amazing manuscript we find that Fibonacci has a version of the three-object divination. His version involves three spectators; one of whom chooses gold, the next silver, the last tin. To determine their choices, Fibonacci assigns them the numbers one, two, and three. The trick now proceeds *numerically* (without coins or counters). The spectator who chose gold doubles his number, the spectator who chose silver multiplies by nine, and the spectator who chose tin multiplies by ten. These three products are added together and the sum is subtracted from sixty. The performer is told the answer to this last subtraction (call it T). The T codes the three choices: Divide by 8 to write $T = 8 \times A + B$. Then A gives the number of the gold chooser, and B gives the number of the silver chooser. This, of course, determines the tin chooser.

We don't quite know what to make of Fibonacci's version. As described, it's a typical, awful math trick, perhaps suitable for a class of school children but far from performable for modern audiences. All of the tricks that Fibonacci describes have this flavor. While it is possible that medieval audiences might have enjoyed this, we doubt it. There is another possibility. Fibonacci was trying to teach people arithmetic, not magic. Maybe he took standard tricks, eliminated presentation and props such as counters, and made them into numerical exercises in a roughly uniform way. If this is the case, it is a worthwhile exercise to go over the other tricks in Fibonacci's book with an eye toward creating performable versions. The three-object divination, the way we have presented it, eliminates any mental arithmetic for the spectators. The counters fade into the background and the effect remembered is one of simple reading. It is certain that some of the old wine will be exquisite if proper serving methods are devised.

Fibonacci presents several technical variations: For instance, the numbers given to the three spectators can be any three consecutive numbers. He develops some theory and says that variations with four or five objects can be devised. This pushed the start of mathematical magic back to 1202. Evidence suggests that, in fact, it may go back much further.

HOW MANY MAGIC TRICKS ARE THERE?

We can use the remarkable coincidence that occurred in 1584 to estimate the number of magic tricks. The coincidence is the appearance of the first two serious magic books, Reginald Scot's *Discoverie of Witchcraft*, and J. Prevost's *La Premiere Partie des Subtiles et Plaisantes Inventions*. As mentioned above, these are very different books. Scot's book is a wide-ranging debunking of witchcraft. Along the way, he gives succinct but clear descriptions of about fifty-two magic tricks. Prevost's book is a "how to perform" manual with longer descriptions of approximately eighty-four tricks.

The two books present us with a natural experiment. We have long felt that the tricks in common usage then don't differ all that much from the tricks in common usage now. At any rate, most of the tricks in Prevost's and Scot's books have a very modern ring—manipulation of coins, cards, and balls, simple mathematical tricks, the cut and restored rope, torn thread, the paddle trick, and many others. Most of these tricks are still widely used today.

Suppose that each author wrote his book by making a more or less random selection out of the common pool of tricks in general use at that time. Suppose there are n tricks in common use, of which Scot chooses s and Prevost chooses p. Then one can estimate n from a count of the number c of common tricks. This procedure is known as a "capture/recapture" estimate; it is commonly used to estimate the number of fish in a lake—take a sample of size s, tag them, then take another sample of size p, and count the number c of tagged fish in the second sample. The method was used to estimate the number of people not counted in the U. S. census in the year 2000.

The usual estimate of n is

$$\hat{n} = \frac{(s+1)(p+1)}{(c+1)}.$$

In the magic books example, $s = 52$, $p = 84$, and $c = 7$. This gives the estimate $\hat{n} = \frac{53 \cdot 85}{8} \approx 563$ for the number of tricks in common use in 1584.

This seems high at first. To put it in perspective, consider that the two books together contain about 129 tricks between them. Scot's book contains only a handful of card tricks: A brief description of card control with a fairly modern ring to it and the marvelous trick involving cutting to the aces and changing them to kings. Prevost's book contains no card tricks at all. It doesn't seem unreasonable to posit 100–200 card tricks in the common domain. After all, Gerolamo Cardano (1501–1576), a celebrated writer, physician, and mathematician, describes modern-sounding card tricks much earlier, and books containing dozens of card tricks appear soon thereafter.[7]

Conversely, Prevost's book contains a handful of mathematical tricks, while Scot's book contains none. The next French book to appear (Bachet's) is filled with mathematical tricks. It doesn't seem unreasonable to hypothesize 75–100 mathematical tricks then in common usage. Neither book explicitly describes the cups and balls or several other tricks that are written up soon after. Even this rough estimate gives about $129 + 200 + 150 = 479$ tricks. We hope this makes the estimate of 552 tricks more plausible.

The method of estimating $\hat{n} = \frac{(s+1)(p+1)}{(c+1)}$ comes along with a standard estimate of its accuracy. One shouldn't hope for an accurate estimation based on two samples. The usual estimate of the standard error $\hat{\sigma}$ for $\hat{n}$ is:

$$\hat{\sigma} = \sqrt{\frac{(s+1)(p+1)(s-c)(p-c)}{(c+1)^2(c+2)}}.$$

In the Scot/Prevost example this gives $\hat{\sigma} \approx 163$. A classical confidence interval for n is $\hat{n} \pm 1.645\hat{\sigma}$. This gives the interval $[234, 820]$ centered at 552 as a 90 percent confidence interval for the number of tricks in common usage.

For the above calculation to be valid, it is necessary to think of at least one of the two authors making a random selection of the tricks from a common pool. This is a bit far-fetched—one may imagine an author as more likely to include popular tricks or simpler, easy-to-perform or easy-to-describe tricks. Any such deviation from a random sample leads to an increase from the above method of estimating $\hat{n}$ by n. If there were a strong deviation in one of these directions, it would lead to a large overlap. We find the overlap of seven remarkably small. For the record, we counted the following seven tricks in the overlap: Grandmother's Necklace, The Box of Grain Transposition, Burnt Thread, Blow Book, Eating a Knife, Knife through Tongue, and Cut-Off Nose. Of course, if the books were aimed at different audiences, then the potential overlap is diminished.

Let us finish these caveats by recalling that later books did a huge amount of copying from these first sources. It seems clear that Scot's and Prevost's books were independent efforts, while later books weren't. Thus, these first books present us with a unique opportunity to do some educated detective work. Background on capture/recapture methods may be found in a survey article by George Seber in the *Encyclopedia of Statistics*.[8]

We began this chapter by telling the story of how one of your authors became interested in mathematical magic and then in mathematics. Here is a similar "crucial moment" for your second author. A young boy of twelve sits quietly in the back of his seventh-grade algebra class, idly looking out the window. The teacher eventually notices that the boy

HARRY ELLS junior high school was the scene of a return chess match recently between the California School for the Blind and the Harry Ells chess team. Shown here during the first game of the match are, seated, Reece Griffith of the Blind school, and Ronnie Graham, of the local team. Standing are the team coaches, Harry Kipps and Richard Schwab.

—The Richmond Independent

Figure 2. Richard Schwab and one of your authors (circa 1946, from the *Richmond Independent*)

is not paying attention to the material, and after class asks him what's wrong. The boy eventually admits that he already knows how to solve all the problems in the class book and frankly wonders if there are any math problems that he *couldn't* solve. The teacher thinks for a minute and then says, "Here is a problem I don't think you can do. Imagine that you start with a population of one hundred mice that then begin to breed. However, the *rate* at which they breed is proportional to the *square root* of the number of mice that are currently present. Thus, as the number of mice increases, the rate at which they multiply also increases. The question is then how many mice will there be at some fixed time later." Well, the teacher was right. The boy couldn't solve a problem like this (in particular, because the solution involves solving a differential equation, something he had never even heard of before). The sympathetic teacher then handed the boy a book and said, "Read this book. By the time you come to the end, you will be able to solve problems like the 'mice' problem."

The teacher was Richard Schwab at the Harry Ells Junior High School in Richmond, California, and the book was *Elements of the Differential and Integral Calculus* by Granville, Smith, and Longley, published in 1941. (Now available on Amazon for as little as twenty-five cents!) To the boy at that time, the book was magical. Beautiful trigonometric formulas, mysterious derivatives, and amazing integrals, all linked together by these elusive, infinitely tiny dx's that went racing down to zero. It was simply an amazing eye-opener at a critical time in the boy's development that had a profound effect on his life, even to this day.

Chapter 8

MAGIC IN THE *BOOK OF CHANGES*

Mathematicians are sometimes seen as the ultimate nerds. An old joke goes: "An outgoing mathematician is one who looks at *your* shoes during conversation." Your authors do not fall into this mode. Each of us has had show business careers and gives more than fifty talks a year (in addition to our scheduled classes). By now, nothing makes us nervous in public presentations. Except for just one time!

In May 1990, John Solt invited us to give a talk at Harvard's Department of East Asian Languages and Civilizations colloquium. We are not sinologists but knew John through magician-historian Ricky Jay. We had been fiddling with some probability and magic speculations around an ancient Chinese text, the *I Ching*, or *Book of Changes*. The idea of a talk emerged as a way of making contact with the community of Chinese historians who occupy the Harvard-Yenching Institute. We brazenly gave our talk the title:

Secrets of the *I Ching* Revealed.

We were surprised to find an audience of fifty or so, from blue-haired little old ladies who use the *I Ching* for fortune-telling to graduate students who study the book as one of the five canons of Chinese literature. It turns out that the *I Ching* is about three thousand years old and, in Chinese culture, is a rough equivalent of the Old Testament in Western culture. Some of our Chinese colleagues in Harvard's Mathematics Department had to memorize sections of the *I Ching* in grade school. They were in the audience. Most intimidating was a group of professors who had just run a six-month seminar on the *I Ching*. Some of this appeared in the book *Sung Dynasty Uses of the "I*

Ching." This traces the appearance and influence of the *I Ching* in the work of four great figures of the Song Dynasty (AD 960–1279). These professors were *very* skeptical about the potential contributions of two mathematicians/magicians.

We had not expected any of this. We were nervous, and rightly so. There is a huge language gap between historians and mathematicians. Further, the idea of doing magic tricks using the *I Ching* was offensive to some. Nonetheless, it turned out that we did have some new things to say. As explained below, the standard method of using the *Book of Changes* for divination involves probability (randomization) developed thousands of years before probability was understood in the West. The talk wound up going well and has been repeated several times (see figure 1).

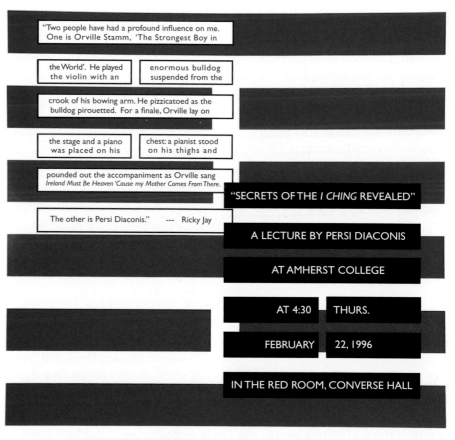

Figure 1. The *I Ching* poster

INTRODUCTION TO THE *BOOK OF CHANGES*

At the heart of the *I Ching* are sixty-four combinatorial patterns, called *hexagrams*. Each has six lines and each of the lines can be *straight* —— or *broken* — —. For six lines, with each being one of two types, this makes $2 \times 2 \times 2 \times 2 \times 2 \times 2 = 64$ patterns in all. All sixty-four combinations are shown in figure 2.

Each hexagram has a name (e.g., Hexagram 1 is "The Creative") and a set of brief comments. These comments have been written by sages such as Confucius. The writing is nonlinear and filled with images. For example, under "The Creative" the text begins: "The creative works sublime success, furthering through perseverance. The movement of heaven is full of power. Thus the superior man makes himself strong

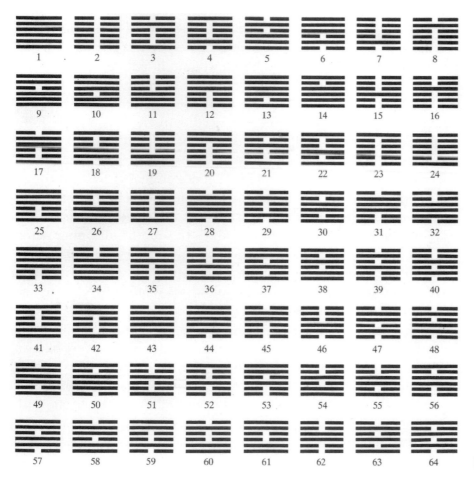

Figure 2. The sixty-four *I Ching* symbols

and untiring. . . ."[1] The hexagrams and commentaries are arranged in separate small sections and make up the body of the *Book of Changes.* (See, for example, K. Smith et al., *Sung Dynasty Uses of the "I Ching."*[2])

The patterns and commentaries have been studied as philosophy and literature for close to three thousand years. The *I Ching* is woven into Taoist and Buddhist religions and its images appear reflected again and again in so many ways that it is impossible to understand Chinese intellectual history without a thorough grounding in the *I Ching.* This has spawned an enormous body of commentary.

USING THE *I CHING* FOR DIVINATION

The *I Ching* is often used, in the West and the East, for divination, to ask about the future. Briefly, one comes to the book with a question. Then, a hexagram is generated "at random." The hexagram is looked up in the book and the commentary is interpreted as an answer. In our college years, fortune telling with the *I Ching* was popular. We had friends who wouldn't leave the house without its advice.

The *I Ching* is sometimes used by artists, composers, and choreographers as a way of choosing the structure of a piece or to trigger creativity. One well-known adherent was the American composer John Cage. Cage's music is challenging to most of us but it makes you think about how the ear and brain construct "music." We knew Cage in the 1960s and vividly recall him taking us on a "listening walk" through New York's Greenwich Village. It was about 2:00 a.m., a heavy rain had stopped, and he grabbed an arm and said, "Come on outside and listen to the city." New York has all kinds of sounds—dripping water, the clicking of traffic lights, the hiss of steam coming up through manholes in the street, taxis, garbage trucks, people's late-night laughter. . . . To this day, if we are stuck somewhere waiting, we can change our mindset and just sit there and listen.

Cage's music is still jarring to most people. One of your authors can offer a conflicting report. A sibling of his was a concert pianist. He had to perform Cage's Sonata for Prepared Piano. This author heard the strange piece, with its fits and starts, erasers and pencils stuck between piano keys, and so on, practiced for weeks. In the end, it became music for the listener. If he hears it today, fifty years later, he gets that happy "Aha" feeling, the same as he does when hearing a more classical piece.

Cage used chance as a compositional device to determine the choice, strength, and duration of various notes. We had an opportunity to revisit these ideas as speakers on the program *The John Cage Legacy: Chance in Music and Mathematics* on November 12, 2008. This included a performance by the Music Committee of the Merce Cunningham Dance Company. Merce Cunningham was there and discussed the use of the *I Ching* in choreography. He explained, "A dancer has two arms, two hands, two feet, and so on. You have to decide where these go and how they change." He said he often used the *I Ching* to trigger these decisions, and allowed that while "Cage took the output of the *I Ching* very seriously and stuck to its choices, I allow myself more fluidity. It's a suggestion. If it doesn't fit, I don't use it." Cunningham and Cage toured the world together performing their dance and music. They used strict rules: If a piece was twenty minutes, Cunningham and his dancers got ten, and Cage and his musicians got ten. The two halves didn't particularly interact. One irate concert

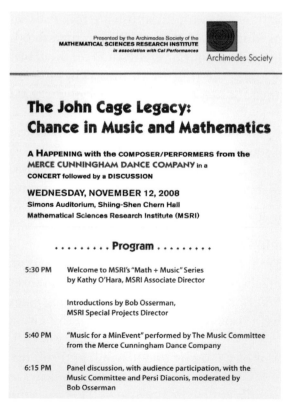

Figure 3. Poster of John Cage talk

goer stormed up afterwards and asked why they didn't integrate, having the dancers dance to the music. Cage answered, "Look. He does his thing, and I do mine; for your convenience, we do them in the same block of time."

Let us return to the *Book of Changes*. The classical way of generating a random hexagram uses forty-nine sticks (nowadays we use pickup sticks but, classically, yarrow stalks were used). These are divided into two piles "at random." One stick is set aside from the right-hand heap. The sticks in the left-hand heap are counted off in groups of four, and the final remainder (1, 2, 3, or 4) is added to the stick set aside. The sticks in the right-hand heap are counted in groups of four, and the remainder (1, 2, 3, or 4) is added to the left pile's remainder. The total number of sticks set aside is

1 + remainder from left + remainder from right.

This will always be five or nine (can the reader see why?). This total, five or nine, constitutes the result of the first step. The total is set aside, the remaining sticks (there will be forty or forty-four) are gathered together, and the procedure is repeated with these (divide into two, set one aside, cast out groups of four in each pile, add the remainders). This second step will result in a pile of four or eight sticks on the table. These are put off to one side, near the first five or nine. The remaining sticks are gathered together and the procedure is repeated a third time (you will again have a remainder of four or eight).

After this, three small piles will have been set aside. Five or nine the first time, four or eight the second and third times. These piles are used to determine a single line of the random hexagram as follows. Use the following rule: 5 and 4 are small, 9 and 8 are large, then set

small = 3, large = 2

for each pile, and, finally, add up the result. Thus, 5, 8, 4 or "small, large, small" yields 3 + 2 + 3 = 8. No matter how the piles are divided, the result of the final sum will always be 6, 7, 8, or 9. These final scores are converted to a single solid or broken line using the changing rules in table 1. Thus, 6 and 8 give a broken line ▬ ▬ while 7 and 9 give a straight line ▬▬.

We will explain the concept of changing and staying, and the displayed probabilities, in a moment. For now, disregard them; if the elaborate procedure above results in 6, write down ▬ ▬. This gives the

Table 1. Probabilities of staying and changing using sticks

				Probability
6	▬ ▬	changes to	▬▬	1/16
7	▬▬	stays as	▬▬	5/16
8	▬ ▬	stays as	▬ ▬	7/16
9	▬▬	changes to	▬ ▬	3/16

first line of a hexagram. To get a complete hexagram, the whole procedure is repeated six times (starting with forty-nine sticks each time). The whole thing can take twenty to thirty minutes. It is part of the ritual of the *I Ching*. One is supposed to think about the question and enter into the ritual. A marvelous description of the effect of this ritual is in Hermann Hesse's description of his hero Knecht's use of the *I Ching* in his novel *Magister Ludi (the Glass Bead Game)*.

To briefly explain the idea of changing lines, begin by generating a random hexagram as explained. After looking it up, one may change all the lines that the table allows to be changed in order to get a new hexagram. This gives a second answer to the question. The commentaries following the hexagrams have a good deal to say about changing lines. The probability calculations that follow 6, 7, 8, and 9 in the table give new insight into changing lines. We turn to these next.

PROBABILITY AND THE *BOOK OF CHANGES*

One of the mysteries in our lives is *why* the study of probability began so late in history. The earliest known systematic probability calculations appear around 1650 in the work of Pascal and Fermat. However, people have gambled in all sorts of ways for thousands of years. There were crooked dice and carefully made near-perfect dice. The ancients used strangely shaped dice made of sheeps' knuckle bones. You would think somebody would have looked at these and said, "Maybe some sides come up more often than others." We do find ample informal discussion of uncertainty as it appears in everyday life. In law, people had to combine evidence from uncertain sources and develop rough rules (for example, two witnesses were better than one, but not if the two were relatives). There were similar developments in medicine and religion. However, we have no record of people calculating odds or having a way to think about randomness other than "the machinations of the gods as seen by mere mortals." James Franklin's magnificent book, *The Science of Conjecture: Evidence and Probability before Pascal*, is surely the best study of the prehistory of probability.[3]

The history of the study of probability in China is similar. In a careful historical study, Mark Elvin of the Australian National University documents the same issues.[4] Gambling was prevalent, and there are comments about uncertainty in law, medicine, and commerce, but *no* calculations or theoretical framework for thinking about uncertainty.

The *I Ching* casts a new light on this mystery. The history of the *I Ching* is not without controversy. Knuth's *The Art of Computer Programming* gives a mathematician's history of the book.[5] We believe the stick method of generating a random hexagram at least goes back to Confucius (551–497 BC). As described above, it uses a complex system of random divisions followed by a summation of "big/small" scores to wind up at a single straight or broken line. The probability calculations summarized in table 1 show that this procedure has been arranged so that the chance of a straight line is $\frac{1}{2} (= \frac{5}{16} + \frac{3}{16})$.

Similarly, the chance of a broken line is $\frac{1}{2} (= \frac{1}{16} + \frac{7}{16})$. The sophisticated procedure isn't obviously symmetric. It takes some thinking for a modern reader to see the 50/50 chance. It took some sophistication to design it in the first place.

We explain where the numbers come from at the end of this chapter. One piece of the argument is worth mentioning now. To do any sort of probability calculations, some assumptions about what it means to "randomly divide a pile of sticks" is required. You can't get probability out without putting probability in. Indeed, not everyone agrees that probability has anything to do with the *I Ching*. During a talk of ours on the *I Ching* at the New York Union Theological Seminary, one of the faculty members angrily protested our probability calculations: "What on earth does probability have to do with the *I Ching*? When I generate a random pattern to use the book by sticks or flipping coins, it is *my* hands that divide the pile or flip the sticks. *I* determine the outcome, not mathematics."

Let us take a serious look at this last complaint. Surely, with practice a longtime user of the *I Ching* can learn to divide a pile with an even number of sticks exactly in two. Indeed, your authors can cut a deck of fifty-two cards exactly in half. Such careful division could even happen subconsciously. If you look back at the full procedure, you will see that randomness in the division is amplified to result in a close to even chance of straight or broken. We describe a mathematical proof of this at the end of this chapter. The ritual seems to have been devised by someone with a feel for probability and combinatorics, thousands of years before such things were clearly understood in the West.

There are other randomization rituals for generating a random hexagram. A fast, frequently used method uses three coins. The coins are shaken in the hands and then tossed on the table. Heads scores three, tails scores two. The sum of all three must be 6, 7, 8, or 9. This is translated into a single *I Ching* line as before. Here, too, the odds are even between ▬ and ▬ ▬, and summing the first and last entries shows that the chance of a changing line is one in four if either coins or sticks are used (see table 2). However, the distribution of hexagrams is quite distorted. For example, a hexagram changing from ▤▤ to ▤ is sixty-four times more likely to occur with coins than with sticks.

A variety of other randomization procedures are in active use, both physically and on the Internet. These can give far different distributions. The stick method has been used for thousands of years, and the coin method for at least a thousand years. Apparently, no one noticed that they were vastly different until the mathematician F. van der Blij carried out the relevant calculations in 1967.[6] We are used to thinking that people learn from experience. Here is a case where they have not.

The *Book of Changes* certainly has a large mystical component. We are complete skeptics about its magical uses. We have often been told of amazing predictions. Anyone who tries out the book will find it answers questions with very, very rich, ill-posed answers. The reader is given a wealth of images and is free to pick, choose, and interpret at will. In addition to the generated hexagram with its commentary, there is the changing hexagram and the relation between the two. Furthermore, each hexagram is made up of two *trigrams*. For example, ▤▤ is composed of ▤▤ and ▤▤. Each trigram has its own name and set of images. A list of some of these appear in figure 5. Thus, if the hexagram "Di" is cast, one might think, "Grace—the mountain is above the sun." What does that mean? In addition to these possibilities, there are several standard arrangements of all sixty-four hexagrams and one may consider one's neighboring hexagrams to get a fuller answer. The arrangements of the sixty-four hexagrams is perhaps the earliest display of binary numbers (interpret ▬ ▬ as zero and ▬ as one). Gottfried Leibniz, Newton's great contemporary, was apparently astounded when someone pointed out that the Chinese had anticipated his discovery of binary numbers by thousands of years. Mysteries remain about these full arrangements.[7]

SOME MAGIC (TRICKS)

With all of this preamble, we hope the reader now has some background on the *Book of Changes*. We next describe three magic tricks

Table 2. Probabilities of staying and changing using coins

				Probability
6	▬ ▬	changes to	▬	1/8
7	▬	stays as	▬	3/8
8	▬ ▬	stays as	▬ ▬	3/8
9	▬	changes to	▬ ▬	1/8

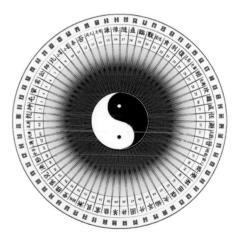

Figure 4. Circular *I Ching* chart (image from *I Ching Meditations* by Adele Aldridge, http://www.ichingmeditations.com)

Triagram	Name	Images	Traits	Family Relations	Parts of Body	Animals
☰	Ch'ien	Heaven Cold	Strong Firm Light	Father	Head	Horse
☷	K'un	Earth Heat	Weak Yielding Dark	Mother	Belly	Ox
☳	Chên	Thunder Spring	Active Moving Arousing	First Son	Foot	Dragon
☵	K'an	Water Moon	Dangerous Difficult Enveloping	Second Son	Ear	Pig
☶	Kên	Mountain	Resting Stubborn Unmoving	Youngest Son	Hand	Dog
☴	Sun	Wind Wood	Gentle Penetrating Flexible	First Daughter	Thigh	Bird
☲	Li	Fire Sun Lightning Summer	Beautiful Depending Clinging	Second Daughter	Eye	Pheasant
☱	Tui	Lake Marsh Rain Autumn	Joyful Satisfied Complacent	Youngest Daughter	Mouth	Sheep

Figure 5. *I Ching* chart 2

based on its ingredients. The first is a fairly old Chinese trick. The second is a modern variation. The third is an elaborate stage presentation. At the very least, from the material above, there is no shortage of surrounding patter themes for the following tricks.

A First Chinese Magic Trick. In effect the performer shows eight pictures or words and invites a spectator to think of any one.

The pictures are on cards that are dealt into two piles. The spectator reports if the thought-of picture is in the left-hand or right-hand pile. This is recorded by a straight ▬ or broken ▬ ▬ line. Repeat this dealing two more times to get a trigram like ☷.

The trick concludes with the resulting pattern actually forming the name (or character) the spectator thought of.

SOME DETAILS. The eight words are shown in figure 6, both as they might be drawn in script and in a more stylized form. The stylized form must be printed on eight cards. Initially, the cards are arranged in a pile in the order 4, 8, 3, 7, 2, 6, 1, 5 from top to bottom.

A spectator is asked to think of any of the eight symbols as they are spread in a wide arc on the table. Put a piece of paper and a pen (or ink and a brush if you want to make a production of this) off to one side. Spread the top four cards, 4, 8, 3, and 7, off and ask if the spectator's thought is contained among these or in the next four, 2, 6, 1, and

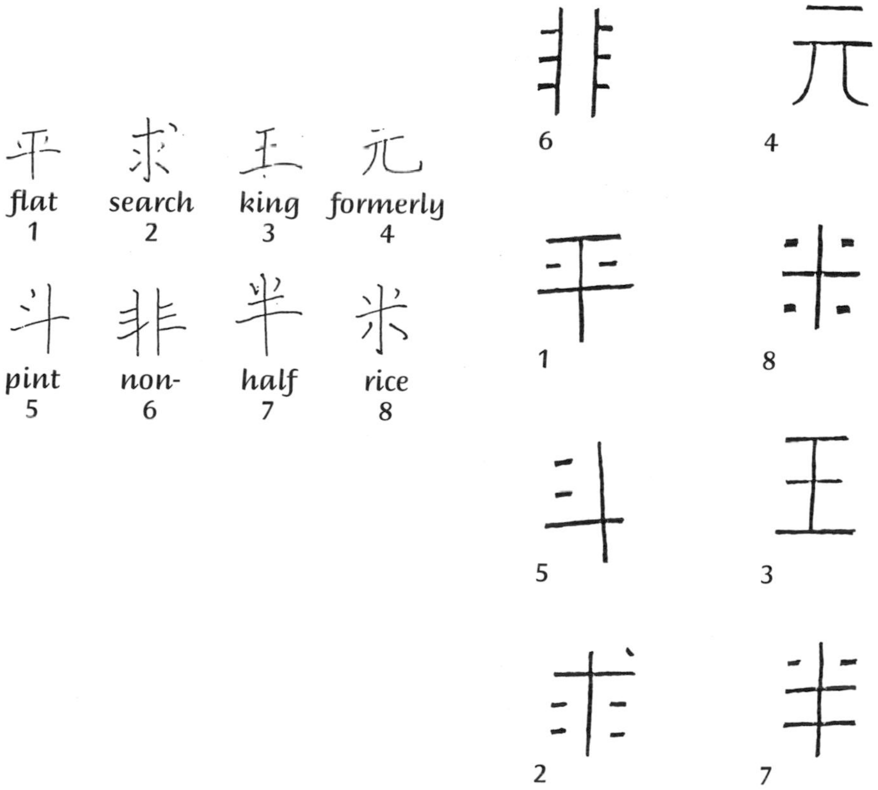

| flat 1 | search 2 | king 3 | formerly 4 |
| pint 5 | non- 6 | half 7 | rice 8 |

Figure 6. *I Ching* chart

5. If the answer is yes for the first four, draw a straight line in the center of the paper ▬. If not, draw a broken line ▬ ▬.

Put the four cards back on top of the packet and deal into two piles, right, left, right, left, etc. Turn over both piles and ask if the spectator's thought is contained in the first pile (1, 2, 3, 4) or in the second pile (5, 6, 7, 8). If it is in the first, make a straight line above the first line drawn. If it is in the second, make a broken line above. Put the first pile on top of the second, deal into two piles, and again show the two piles. If the first pile with cards 1, 3, 5, 7 contains the thought-of picture, make a straight line below the two lines drawn previously. If not, make a broken line.

This procedure results in a trigram of three straight or broken lines. The trick is concluded when the performer adds one final stroke, which changes the trigram into the Chinese spectator's thought-of character.

Here is an example. Suppose the spectator thinks of the word "search" (card 2). At the first stage, this word is not seen so the performer draws a broken line ▬ ▬. At the second stage, this word is seen, so the performer draws a straight line above the initial line ☱. At the third stage, card 2 is not seen, so the performer draws a broken line below the first two ☷. Now, the performer adds a single stroke, forming ☶. This becomes the character for "search." With practice, the final pen stroke plus another pen stroke or two can make the completed picture quite a reasonable likeness of the thought-of character.

We do not know how old this trick is. Our source is a recent Japanese book that calls the trick Su Wu Tending Sheep.[8] Su Wu lived in the Han Dynasty (140–60 BC). He was banished to a remote, poor part of China and spent nineteen years in poverty and near-starvation before being forgiven. There is a still-popular children's song about his difficulties, which also tells of his sheep-tending activities.

The Japanese book says that the trick can be found in an old Chinese book, romanized into pinyin as "Zhōng Wài Xì Fǎ Tú Shuō."[9]

The eight trigrams introduced above form an integral part of the *I Ching*. The chart in figure 4 gives the names and basic images of the eight trigrams as used in the *I Ching*. The last column lists the image used in the trick as explained above. Some of the classical images match the magic trick's words.

A VERSION IN ENGLISH. It is natural to attempt a version of this trick in English. We record two of our attempts here.

THE FIRST VARIATION. The performer shows eight pictures of common objects. A spectator thinks of one. While this is happening, the performer arranges a prediction on a small stand. The spectator reveals the thought to the audience so all can appreciate the denouement, in which the performer shows that the prediction spells out the name of the thought-of picture. Briefly, we found eight different three-letter words with the first, second, and third letters each being one of two choices, e.g., rug, hug, rag, hag, rat, hat, rut, hut. The prediction consists of three pairs of cards lined up on a stand. The first pair has an *R* and an *H*, the second pair has an *A* and a *U*, the third pair has a *G* and a *T*. The pairs are lined up so that they appear as three single cards. The performer can turn over one or both each time and reveal a correct prediction. We have kept the description minimal as it's a poor trick. Let us know if you find a good variation. Incidentally, we were unable to find a good set of sixteen four-letter words with a parallel structure. The best we could find is:

> dice, link, dick, line, dine, lick, dink, lice,
> duce, lunk, duck, lune, dune, luck, dunk, luce.

This isn't wonderful but at least all these words are in most common dictionaries.

THE SECOND VARIATION. This is close to the original Chinese version, but Western images are used for the picture of the thought-of object shown at the end. Here, the sixteen words shown below are displayed on a set of cards, one word per card:

> coffee, goldfish, cowboy, clown, swan, tulip, horse, bunny
> sheep, cake, school bus, house, glasses, trolley, ice cream, mermaid.

A spectator thinks of a word. The cards are dealt into two piles and the spectator indicates in which pile the thought lies. The performer makes a large circle on a pad if the thought lies in the left-hand pile and a large square on the pad otherwise. One pile is put on top of the other and the procedure is repeated three more times. Each time, the performer adds a few lines or squiggles. After the final deal, the performer has finished drawing a picture of the thought-of object. The charming designs shown in figure 7 were created by Laurie Beckett.

HOW IT WORKS. The sixteen words in figure 7 are associated to the binary numbers. Thus, *coffee* is associated to 0 0 0 0, *goldfish* to 0 0 0 1,

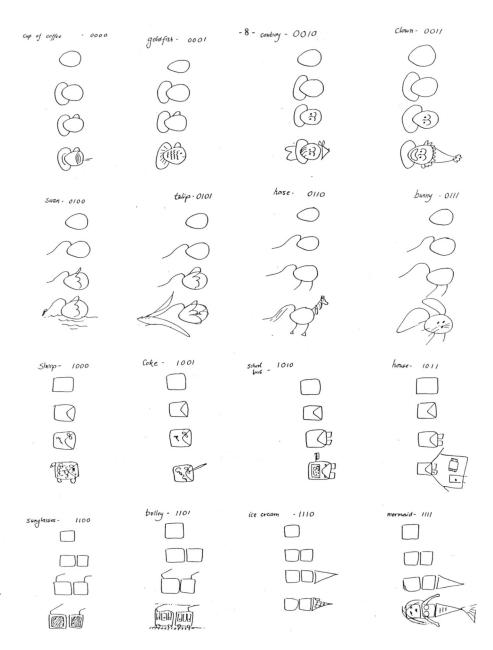

Figure 7. Western images for the *I Ching* trick (courtesy of Laurie Beckett)

and so on, with mermaid assigned to 1 1 1 1. Put the words on sixteen cards and arrange these cards in order, with *coffee* on top and *mermaid* on the bottom.

Spread the top eight cards in one pile and the bottom eight cards in a second pile, so the spectator can see all the words. The cards are spread so that the order is not lost. The spectator thinks of a word and indicates which pile. If it is the original top pile, draw an egg-shaped circle; otherwise, draw a square.

Pick up the piles. Deal the cards into two piles and ask again. A second line is added to your original drawing. The sixteen sequences in figure 7 show which lines are added. After the final deal, add the few relevant details to make the picture appear. This final detailing is just as required with the original Chinese trick.

A PERFORMANCE PIECE. Our third trick using the images of the *Book of Changes* is a performance piece. After giving a little of the background of the *I Ching*, the performer asks for the help of three spectators. A fourth spectator is given an elaborate prediction. Now, the performer asks each spectator to ask a simple personal question that can be answered yes or no. In one recent performance (by P. D. for R. L. G.'s retirement party at Bell Laboratories), one spectator (a Rutgers graduate student) asked, "Will I pass my qualifying exam?" A second spectator (today the most famous computer scientist in China) asked, "Is Fermat's theorem actually proved?" The third spectator (a well-known Hong Kong investor) asked, "Will I make a million dollars this week?" The three questions are recorded at the corners of a large triangle with single-word summaries (see figure 8). For home performances, a large piece of paper is used. In our performance, on a stage before hundreds of spectators, the three helpers were up on stage and the performer drew the triangle on a transparency projector that projected the image on a large screen.

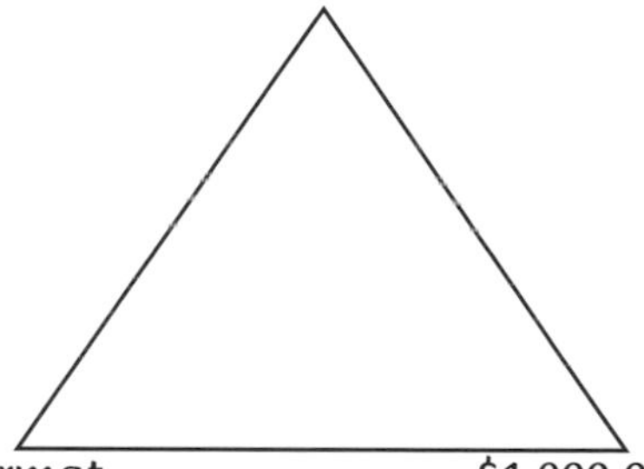

Figure 8. Bell Labs version of *I Ching* trick

The performer now involves the spectators in a simple randomization ritual that leads to a random trigram. In our example, it was ☰, Ch'ien. Explaining that each line of the trigram answers a question, with ▬ meaning yes and ▬ ▬ meaning no, the performer asks the fourth spectator holding the prediction to remove it from the envelope. On one side, the prediction says

☰

The creative, strong, heaven, father, three times yes good omen.

On the other side, are three brief commentaries with a more focused prediction for each. These are:

- *Yes* at the beginning means hidden dragon. Do not act. A low card and the queen reinforce.
- *Yes* at the second place means dragon appearing in the field. It furthers one to see the great man. The men of the pack await.
- *Yes* in the third place means all day long the superior man is creatively active. At nightfall, his mind is still beset with cares. Danger no blame. Wild deuce and joker release you.

As each line is read, the performer relates it briefly to the question asked. Part of the randomization ritual (explained below) involves playing cards. Each of the three spectators winds up with two cards. The prediction not only answers their questions but tells exactly what cards they have.

We have called this a performance piece since it mixes the history and mysticism of the *I Ching* with a magic trick. Depending on the circumstances and audience, we may show the actual book and do a trial hexagram, or talk a bit about probability as explained above. The running time can range from less than ten minutes to nearly a full hour's lecture. It seems to play well for the right kind of audience. It would be a disaster at a noisy party.

The actual randomization ritual involves a packet of fourteen playing cards. In our performance for a large audience, we used extra large giant cards. To follow along, remove the thirteen hearts and the joker from the deck, and arrange them in the order A, 5, 3, 6, 4, J, 2, 8, 9, 10, 7, Q, K, Joker, from top down. These cards will be repeatedly mixed face-up and face-down according to the spectators' directions. The basic mixing step is this: Hold the packet of cards face-down as if you were about to deal them in a card game. Deal four cards into a pile on the table as follows: Deal the top card down, the second face-down on the first, turn the third card over (face-up) and deal it on the first two. Deal the fourth face-down on the first three to make a packet of four on the table. Pick this packet up and turn it over as a block, finally replacing all on top of the big group in your hands. This results in three face-up and one face-down. This basic deal will be repeated many times. We call it a "G-scam" deal. It may seem a bit like the Hummer deals in our first chapter but is actually quite different. Let us finish the *I Ching* trick.

You have three spectators in front of the audience with their questions recorded at the corners of a large triangle. Say that you will harness their input along with some randomization to generate a random pattern. Remove the fourteen playing cards, arranged in order. Carry out a G-scam deal and ask the first spectator to cut the packet at random. Carry out a second G-scam deal and ask the second spectator to cut the packet. Follow a third G-scam deal with a cut by the third spectator. In fact, any number of G-scam deals can be carried out. We find that three is enough.

Explain that an *I Ching* trigram will be formed by using the face-up/face-down pattern in the following manner: an *even* count becomes ━━, an *odd* count becomes ━ ━. Deal the packet of fourteen cards around the large triangle in the following manner. Deal the first three cards at the corners of the triangle in order 1, 2, 3 (see figure 9). Then, place card 4 on the edge of the triangle between card 1 and card 2, place card 5 between cards 2 and 3, card 6 in the center of the triangle, and, finally, card 7 between cards 1 and 3 (see figure 10).

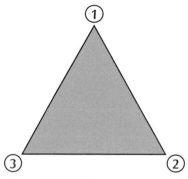

Figure 9. Basic Triangle

The next seven cards are dealt in exactly the same way: The current top card is put on card 1, the next on card 2, the next on card 3, and so on. All of this results in seven piles, each containing two cards.

Explain that the corners give the answers to each spectator's question, written at the corners. For question one, the two-card packets at position 1, its two touching sides, 4 and 7, and the center, 6, are used. Spread these packets out and total up the number of face-up cards showing. If this total is an even number, write ━━. If it is an odd number, write ━ ━. For the second question, use the two-card packets at positions 2, its adjacent sides, 4 and 5, and the center, 6. For the third question, use the two-card packets at positions 3, 5, 6, and 7.

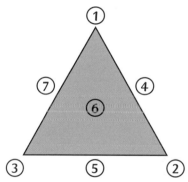

Figure 10. Completed Triangle

The outcome, no matter how many G-scam deals and random cuts are done, will always be ━━. To make the predictions come out right, you must add one additional move. Before the actual dealing into piles starts, the queen *or* four must be on top. To facilitate this, both should be marked with a dot on their backs. If, in handling the cards, you see the queen or four at the bottom, all cards may be turned over. If not, carry out a few extra G-scam deals, and stop when the queen or four are in position. Failing this, make an open cut. This last piece of handling fits right in as the audience doesn't know quite what is coming.

This long description has had many pieces. It will take dedication to make your own performance piece believable. We predict that our

G-scam deals and their generalizations will have a life of their own. We promise the reader that the mathematics involved is just as interesting as the trick.

PROBABILITY AND THE *I CHING*

What are the chances of the various final outcomes (5, 6, 7, 8) if the *I Ching* ritual is performed with a bundle of sticks? To compute these, we need to know the answer to the following question: If a bundle of n sticks is divided into two piles at random, what are the chances that the left-hand pile has j sticks? Here, j can run through anything from zero to n, although we expect the sizes of the two piles to be roughly the same.

We thus arrive at the question: What is the appropriate mathematical model for the random division of a pile of n sticks? The famous French mathematician Laplace considered the uniform distribution as basic. Here, all divisions are equally likely. With forty-nine sticks, the chance of j on the left is $\frac{1}{50}$ for j between zero and forty-nine. A natural alternative model is the binomial distribution. This results in much more even piles: The chance of j on the left is given by the formula $\binom{n}{j}/2^n$. When $n = 49$, the chances are shown in table 3. We see that the chance of an even split (24, 25) is 0.1123, much larger than $\frac{1}{50} = 0.02$ in the uniform model.

Which model is correct, and does it matter? These are empirical questions, amenable to experiment. While this would be instructive, it can be proved that the exact distribution is irrelevant to a good approximation. The point is, the exact chance of j doesn't matter, but rather just the chance that j has a given remainder (0, 1, 2, 3) when divided by 4, and these chances are very close to $\frac{1}{4}$ for a wide variety of probability distributions. For example, for the uniform and binomial distributions, the chances are shown in table 4. A much more general result can be proved.[10]

Table 3. Probability that j sticks end up on the left

j	Probability that j sticks end up on the left
20	0.0502
21	0.0694
22	0.0883
23	0.1036
24	0.1123
25	0.1123
26	0.1036
27	0.0883
28	0.0694
29	0.0502

Table 4. Probability of j mod four for different distributions

j mod 4	Binomial	Uniform
0	0.2600000000	0.2500000149
1	0.2600000000	0.2500000149
2	0.2400000000	0.2499999851
3	0.2400000000	0.2499999851

WHAT GOES UP MUST COME DOWN

Juggling, like magic, has a long history, both dating back at least four thousand years. Indeed, magic and juggling are often associated with each other. Certainly, some of the top jugglers seem to have supernatural abilities, while a number of magic tricks (such as perfect shuffles) require highly developed physical skills. In fact, many talented magicians are also skilled jugglers, e.g., Ricky Jay and Penn Jillette. Moreover, there is also a strong connection between juggling and mathematics. Mathematics is often described as the science of patterns. Juggling can be thought of as the art of controlling patterns in time and space. Both activities offer unbounded challenges. In mathematics, you can never solve *all* of the problems. In juggling, you can never master *all* the tricks (just add one more ball!). In this chapter we explain one real connection between mathematics and juggling. Keeping with our tradition of encouraging performance, at the end we teach the basic "three-ball cascade."

Figure 1. An image from the fifteenth Beni Hassan tomb of an unknown prince, from 1994 BC

WRITING IT DOWN

During the past two decades a remarkable connection between mathematics and juggling has come to light. This commonly goes under the name "siteswaps." A *siteswap* juggling sequence (or pattern) is just a (finite) sequence of numbers that specify the amount of time the thrown objects are in the air. For example, 534, 4413, and 55514 are siteswap patterns. We'll explain how these represent juggling patterns in what follows. We will use balls as our objects to juggle (although more challenging objects can be juggled, such as fire torches or chain saws, the theory is the same). We imagine that time moves along in time steps 1, 2, 3, . . . (which we can think of as seconds; see figure 2).

Figure 2. Time steps

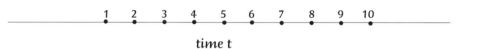

Now, suppose at time 1 we throw a ball with time value 3. This means that this ball will come down at time 4 = 1 + 3 (see figure 3). If we were to represent this as a siteswap sequence, it would be 3000. . . . Ordinarily, siteswap notation is designed to represent repeating patterns, so this particular representation of a single throw is not so useful.

Figure 3. An easy trick

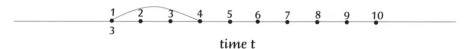

Of course, a single throw is not a very impressive trick. More interesting is the siteswap juggling sequence 333333 . . . , which we will abbreviate as 3. For this pattern, as each ball lands, it is immediately thrown back up into the air with a time value 3. We can diagram this pattern as shown in figure 4.

Figure 4. The siteswap 3333333 . . . = 3

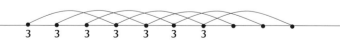

In general, a siteswap pattern consists of a sequence of numbers t_1 $t_2 \ldots t_n$, which are all greater than (or equal to) zero. The interpretation is that at time i a ball is to be thrown so that it lands at time $i + t_i$. The usual interpretation for siteswap patterns is that they are performed by a two-handed juggler with the left and right hands throwing alternately. Thus, if we put R and L in the diagram to denote which hand is throwing the ball at that time, we have the picture in figure 5. This is the siteswap notation for the basic three-ball cascade (which some readers may already be able to do). Notice that if a throw value is *odd* (such as for the siteswap 3 = 3333333 . . .) then one hand throws the ball and the *other* hand catches it. In contrast to this, for an *even* throw, the same hand throws and catches the ball (see figure 5).

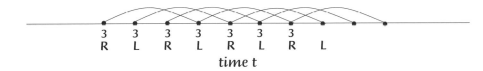

3   3   3   3   3   3   3
R   L   R   L   R   L   R   L
time t

Figure 5. Parity considerations

This is an example of *parity* in juggling, i.e., a fancy way of distinguishing between even and odd numbers. We point out that from a physiological perspective, somewhat different feedback circuits through the brain are used for these different types of throws. Some jugglers are better at the "crossing" patterns (where the balls change hands), and some are better with patterns in which the balls don't change hands.

One of the first siteswap patterns that people learn is the pattern 441441441 . . . , which we shorten to 441 (see figure 6). It has a very nice rhythm to it, but is a bit harder to do than it looks, since the "1" throws in the pattern involve throwing balls straight across both from the right hand to the left hand, *and* from the left hand to the right hand, something that takes a little practice to do consistently.

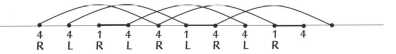

4   4   1   4   4   1   4   4   1   4
R   L   R   L   R   L   R   L   R

Figure 6. The siteswap 441

Another interesting pattern is 534534534 . . . = 534 (see figure 7). The pattern 534 is definitely more challenging to juggle than 441. Notice that in this pattern, no two balls come down at the same time.

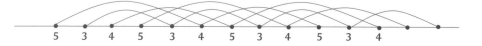

Figure 7. The siteswap 534

In general, all our patterns will be repeating. The number of throws before it begins repeating is called the *period* of the pattern. We usually just write down one cycle of the pattern.

One could start writing all kinds of potential siteswap sequences (and try to juggle them). For example, what about 543 (see figure

Figure 8. The potential siteswap 543

8)? Well, you see we have a problem, namely, several (in this case, three!) balls come down at the same time to the same hand (jugglers hate when this happens). How can we tell ahead of time if this is going to happen? If $(t_1, t_2, \ldots, t_n)$ is our juggling sequence then this means that the ball thrown at time 1 lands at time $1 + t_1$, the ball thrown at time 2 lands at time $2 + t_2$, and, in general, the ball thrown at time i lands at time $i + t_i$. Thus, after one pass through the period, the balls land at times $1 + t_1, 2 + t_2, \ldots, n + t_n$. So, in particular, all the quantities $i + t_i$ for $i = 1, 2, \ldots, n$ better be *different*. However, this isn't quite enough to guarantee that we won't have a collision problem. For instance, consider the pattern 346 (see figure 9). We can check that $1 + 3 = 4$, $2 + 4 = 6$, $3 + 6 = 9$ are all different, so are we okay? Well, remember 346 stands for the *periodic* repeating pattern 346346346. . . . The period of this pattern, of course, is 3. Thus, at times 2, 5, 8, . . . , we throw a 4, and they come down at times $2 + 4 = 6$, $5 + 4 = 9$, $8 + 4 = 12$, etc. Similarly, at times 3, 6, 9, . . . , we throw a 6, and these come down at times $3 + 6 = 9$, $6 + 6 = 12$, $9 + 6 = 15$, etc. So, in particular, we have a collision at time 12 (and also at times 15, 18, . . .). We can see this if we draw in a few more of the throws, as shown in figure 10.

Figure 9. The pattern 346

3   4   6   3   4   6

Figure 10. A collision in the pattern 346346 . . .

In a general juggling sequence, what we need to avoid this is for all the quantities $i + t_i$, $i = 1, 2, \ldots, n$ to be different *modulo n*. In other words, we need all the remainders of the $i + t_i$ after subtracting off the largest multiple of n to be different. For example, for 441, the period is 3 and the remainders of $1 + 4$, $2 + 4$, and $3 + 4$ are 2, 0, and 1, respectively, so this pattern is "juggleable," i.e., a valid siteswap.

Similarly, for 534, the period is also 3 and the remainders of $1 + 5$, $2 + 3$, and $3 + 4$ are 0, 2, and 1, respectively. However, for 543, the sums and remainders are $1 + 5$, $2 + 4$, $3 + 3$, and 0, 0, 0, respectively, which explains why three balls came crashing down together as we saw earlier (which is a catastrophe!).

This is the first theorem for siteswap patterns:

> THEOREM. A sequence $(t_1, t_2, \ldots, t_n)$ with all $t_i \geq 0$ is a valid siteswap pattern (or, is juggleable) if and only if all the quantities $i + t_i$, $i = 1, 2, \ldots, n$ are distinct modulo n. This theorem reminds us of our Ultimate Gilbreath Principle in chapter 5. Is there any connection?

We remark that the value $t_i = 0$ just means that no ball is thrown (or caught) at time i. If one has a siteswap pattern $= (t_1, t_2, \ldots, t_n)$, it is important (for a juggler, anyway) to know just how many balls are needed to juggle it. For example, how many balls are used in the pattern 534? (Follow the bouncing balls in figure 7.) For this pattern, the answer is four. On the other hand, for the pattern 441, the answer is three. It turns out there is a very neat expression for the number b of balls there are in the pattern $(t_1, t_2, \ldots, t_n)$. It is just the *average* of the t_i, i.e.,

$$b = \frac{1}{n}(t_1 + t_2 + \ldots + t_n).$$

(Check the patterns 534 and 441.)

One way to see this is to imagine that, at each time step, one ball goes up and one comes down. The average amount of time a ball is in the air is $\frac{1}{n}(t_1 + t_2 + \cdots + t_n)$ time steps. Hence, there needs to be exactly this many balls! (Note: This is not exactly a mathematical proof; in fact, it is more of a hand-waving proof!)

Ah, you might say, this is nice but what if this quantity turned out not to be a whole number (such as $\frac{9}{4}$)? That would put a bit of a crimp into this theory, would it not? Indeed it would, but don't worry, it can never happen. Here's why.

By hypothesis, for a valid siteswap $(t_1, t_2, \cdots, t_n)$ all the quantities $i + t_i$, $i = 1, 2, \cdots, n$ must be distinct modulo n. Since all the remainders after dividing by n are between 0 and $n-1$, and there are n of them, then, in fact, they must be exactly all the quantities $0, 1, 2, \ldots, n-1$ in some order. Hence, if we add them all up, we just get the sum $(0 + 1 + 2 + \cdots n - 1)$ modulo n. That is,

$$(1+t_1) + (2+t_2) + \cdots + (n+t_n) = (0+1+2+\cdots+n-1) \quad \text{modulo } n.$$

However,

$$(1+t_1) + (2+t_2) + \cdots + (n+t_n) = (0+1+2+\cdots+n-1)$$
$$+ (t_1+t_2+\cdots t_n) \quad \text{modulo } n.$$

Consequently, subtracting we get

$$t_1+t_2+\cdots t_n = 0 \quad \text{modulo } n,$$

i.e., $t_1+t_2+\cdots t_n$ is divisible by n. This implies that the number $b = \frac{1}{n}(t_1+t_2+\cdots+t_n)$ is a whole number (whew!). However, this average can also be a whole number for some invalid sequences as well, such as the sequence 543 we saw earlier.

The next question a mathematical juggler might ask is just *how many* different siteswap juggling patterns there are with period n and b balls. It turns out that there are a lot of them. The exact number is given by the slick expression

$$(b+1)^n - b^n.$$

Thus, for $b = 4$ and $n = 3$, there are $5^3 - 4^3 = 61$ different possible four-ball siteswaps with period 3. (The sequence 534 is one of them. Can you list the other sixty? Can you juggle some of them?) Strictly speaking, this is an overcount, since we include in this number any pattern with a period dividing n. For example, for $b = 4$ and $n = 3$ we are also counting 444, which has period 1 and can be written more succinctly as 4.[1]

Once the connection has been made between juggling (sequences) and mathematics, all kinds of doors, both mathematical as well as juggling, are thrown wide open. Many jugglers have been working hard to master the almost unlimited number of new patterns suggested by siteswaps. This includes the generalizations to doing siteswap *passing* patterns between two or more jugglers, as well as *multiplex* juggling patterns, where more than one ball may be caught and thrown from an individual hand. Many of the siteswap patterns are difficult. For example, contrast a standard juggling pattern of a five-ball cascade (described as 5 = 5555 . . . in siteswap notation) with a five-ball siteswap such as 4637. In the first pattern, each ball gets thrown at the same height and to the opposite hand, whereas, in the second, the balls go to all different heights, some change hands, and some don't. (We have never actually seen anyone juggle this pattern.)

On the mathematical side, there still remain many interesting challenges. For example, as we have seen, it is necessary for a valid siteswap pattern $(t_1, t_2, \ldots, t_n)$ that the average $\frac{1}{n}(t_1 + t_2 + \cdots + t_n)$ must be a whole number (although this is not sufficient, as 543 shows). However, it can be shown that, for any sequence $u_1, u_2, \ldots, u_n$ of n numbers for which the average is a whole number, it is always possible to *rearrange* them to get a valid siteswap pattern. For example, 534 is a valid rearrangement of 543. The pattern 252505467 has period 9 and a sum of 36 (and an average of 4) so there must be at least one rearrangement of these numbers that is a valid siteswap. (Can the reader find one?) The proof that this can always be done in general is very tricky. If any reader can find a nice way to show this please let us know! Incidentally, no one knows which sequences have the *largest number* of rearrangements that are siteswap sequences.

The whole approach to juggling patterns through siteswaps can be extended to *multiplex* juggling patterns as well. In this case more than one ball can be thrown and caught at any particular time. Naturally, this is harder for a juggler to do (but many jugglers are quite good at it by now). These same concepts can also be applied to groups of jugglers who exchange balls (or, more commonly, juggling clubs) while they are juggling. In these cases, the mathematics and the juggling both get more sophisticated. In fact, a number of new mathematical ideas have come out of these considerations, such as new ways for computing the so-called Poincaré series for the affine Weyl group, *q*-nomial Stirling number identities, new Eulerian number identities, and enumerating certain rook polynomial configurations.[2] Going in the other direction, (expert) jugglers are now trying to master some of the more mathematically interesting siteswap patterns, such as 123456789. In fact, if you make the associations a ↔ 10, b ↔ 11, etc., then one might ask which (English) *words* can be juggled. It turns out that the words "theorem" and "proof" are both juggleable patterns, the first being a twenty-one-ball pattern, while the second is a twenty-three-ball pattern. Well, we always knew that proofs were harder than theorems!

There are a number of siteswap simulators for computers available on the Web, which allow you to input a potential siteswap pattern and watch it being performed on your computer screen. This is an especially good way to learn some of the more complex patterns. The most popular one is probably Jongl.[3]

One of the problems in juggling is that gravity is inconveniently strong on the Earth. It has been estimated that a seven-ball juggler on the Earth could juggle forty-one balls on the moon (which has gravity one-sixth as strong as the Earth), except for certain practical difficulties, such as being in a spacesuit and the fact that the pattern doesn't "scale," i.e., if you throw a forty-one-ball pattern forty feet up, you have to be very accurate since the spread between your two hands is still only (at most) six feet. One way to slow gravity down is to roll balls on a (slightly) inclined table, or bounce them off the cushions of a billiard table (in which case, the paths made by a single throw become very "pointy" at the top). You can also bounce silicone balls on a smooth surface, such as down to the floor, horizontally off a wall, or, in the case of the famous American juggler Michael Moschen, on the three sides of a large triangular enclosure, all of which gives the juggler more time between throws.

One of your authors got seduced into the world of juggling as a teenager, fascinated by the permutations and combinations that seemed possible. He even managed to create some new "entangled" patterns based on combinatorial ideas (like the Mills Mess, perfected and popularized by his student Steve Mills of the highly successful Dazzling Mills Family juggling act). He subsequently served as President of the International Jugglers' Association, a group of some three thousand active jugglers around the world (with quite a few having day jobs in computing, mathematics, and the sciences generally). Most large cities and college campuses have active juggling clubs. A listing of these, including when and where they meet, as well as almost any other juggling need, such as props, videos, books, newsletters, meetings, competitions, performer itineraries, etc., can be found on the Juggling Information Service Web site.[4]

As they say, "Old jugglers never die, they just lose their . . . props"!

GETTING STARTED IN JUGGLING

It is not as hard as you might think to learn some of the basic three-ball juggling patterns. In the next few paragraphs we will present a brief tutorial on how to do it.

To begin, it is necessary to have some reasonable objects to juggle! Typically, some kind of ball works best, say about the size of a tennis ball, but heavier, if possible. Many beginning jugglers use lacrosse balls or "dog" balls from the local pet shop. Tennis balls can work but they are too light to be optimal. Some jugglers fill tennis balls with sand or some kind of grain to make them heavier, but then they don't bounce much (which, at the beginning, is a good thing!). Golf balls are too small, volleyballs are too big, and footballs are wrong for a variety of reasons. Recently, the most popular objects for juggling are beanbags, which are available from a number of sources.[5]

STEP ONE. There are basically three steps in learning what is called the three-ball cascade, the first pattern most jugglers learn. Step one involves just a single ball. You start with a ball in one hand, with both hands held at about waist level. You will then throw the ball up into the air and across to the other hand. The ball should go up to about six to twelve inches above head height, and you should follow the ball with your eyes for most of the time it is in the air (see figures 11–14).

You should resist the temptation to reach up and grab the ball while it is still head high. It will definitely come down by itself! You should concentrate on trying to throw the ball so that it stays roughly the same distance from your body, and doesn't shoot out away from you in front or come back and hit your chest. A helpful technique is to stand about a foot from a wall and try to keep the path of the ball the same distance from the wall. In general, it is helpful to practice with a nondistracting background, such as a solid wall, in front of you. This practice should be done with each hand as the starting hand. By all means, resist the impulse to hand the ball from one hand to the other *without* throwing it up in the air. Many people instinctively want to throw the ball with their dominant hand (typically the right hand) and then, after catching it in the other hand, *handing* the ball back to the starting hand without throwing it. For this reason, it is useful to

Figure 11. One ball starting position

Figure 12. Throwing one ball

Figure 13. Ball is at the top of its arc

Figure 14. Watching the ball on the way down

practice the left-hand to right-hand throw (for right-handed people) somewhat more than the other direction until both directions feel comfortable. This process usually takes from three to five minutes.

STEP TWO, which is the most important, involves only two balls. For this step, you start with a ball in each hand. We'll assume for the moment that you are right-handed. (If not, you should reverse the instructions.) This is what you will try to do. You will throw the ball in the left hand up and across to the right hand, as you were doing in step one. However, *as this ball reaches its peak*, you will then throw the ball in the right hand, *up* and across to the left, making sure that it passes *under* the first ball (see figures 15–22).

At first, it may seem that there isn't enough time to throw the second ball, but in a minute or two you'll find that this isn't a problem. There are several things you should try to do for this step. First, make sure you wait until the first ball has reached its peak before you throw the second ball (see figures 18 and 19).

They should definitely *not* be thrown at the same time! Also, it is important that you watch the first ball until it is almost caught, and

Figure 15. Two balls starting position

Figure 16. Throwing the first ball

Figure 17. Watching the first ball

Figure 18. Preparing to throw the second ball

Figure 19. Second ball thrown and first ball caught

Figure 20. Watching the second ball

Figure 21. Still watching the second ball   Figure 22. Second ball caught

then switch your focus to the second ball. After all, you are going to have to catch the first thrown ball first, so you better be sure that you can catch it before worrying about the second ball. If you find that you are always dropping the first thrown ball, it is a sure sign that you are looking away from it too soon. Also, make sure that the second ball goes up just as high as the first ball. There is a common tendency to throw the second ball significantly lower than the first ball, which will cause problems for step three. Finally, as in step one, try to keep the balls in the same plane, so you don't have to reach way out in front or back to catch them. Remember, this step only has two throws. We do not continue throwing more balls at this point.

Once this two-ball exercise can be done reliably (say nine out of ten times), try to do the same thing starting with the other hand first. You will probably find that this is more awkward than you expect but you should go through the same steps as you did before. Fairly soon, you will be able to do either of the two-throw patterns with relative ease. This whole process typically takes fifteen to thirty minutes, with very talented people taking under ten minutes and those less talented taking up to an hour. You will find that you are using more energy than you realize during this step, so you might want to take a break for a few minutes after twenty minutes and grab a drink (of water!).

STEP TWO-AND-A-HALF (okay, we lied) really is only a half step. What you do here is execute step two but *hold a third ball in your right hand*. The ball should be held in the back part of the hand, with the thrown ball coming from the front of the hand (see figure 23). In this step, we *don't* throw this third ball. We only throw two balls—first one from the right hand and then one from the left hand (as in step two) but then, as the second thrown ball is coming down to the right hand, *think* about the possibility of throwing the third ball under the landing second ball. At first you probably won't have quite enough time to actually throw the third ball, but thinking about it will make you aware of the possibility. As before, during this half step, keep the thrown balls at the same height (because if the second ball is thrown a lot lower than the first ball then you definitely won't have enough time to throw the third ball!), wait until the first ball reaches its peak before throwing the second ball, and keep the balls in a plane in front of you.

STEP THREE. As you are executing step two-and-a-half, you will see that it *is* possible to throw the third ball under the second ball from the right hand to the left hand—just do it! This will (soon) happen naturally since in step two you have already practiced throwing the left hand first, and that is all that the second and third balls are doing. Don't rush. Count to yourself: right, left, right, waiting for each ball to reach its apex before throwing the next ball. Give each ball enough attention while you are throwing it in a proper trajectory (good height, plane, and timing) and when you are catching it. If you have mastered step two (and step two-and-a-half) you should find that within five minutes you will be able to perform step three, catching all three balls. At this point, you should congratulate yourself. You have achieved what jugglers call a "flash" of a three-ball cascade.

STEP FOUR AND BEYOND. Once you can do step three and, with some confidence, flash the three-ball cascade, it is a simple matter to imagine when the *third* ball is about to land in the left hand, that it *might* just be possible to throw the ball already in that hand over to the right hand before the third ball lands. Again, this is just one more occurrence of step two. Think about it a few times, and then do it. Once

Figure 23. Three balls starting position

Figure 24. Throwing the first ball

Figure 25. Preparing to throw the second ball

Figure 26. Second ball thrown and first ball caught.

Figure 27. Watching the second ball and starting to throw the third ball

Figure 28. Second ball caught and third ball thrown

Figure 29. Watching the third ball

Figure 30. Third ball caught—a three-ball cascade flash

that is accomplished, you will begin to see the pattern. In mathematics, this is called "induction," namely, if you know how to go from step n to step $n+1$ then you can keep going indefinitely. You see, there is mathematics in juggling.

Your first goal at this point might be to achieve ten throws in a three-ball cascade. If you manage this after one hour, you are a great student. If you are more ambitious, try for twenty-five. If you find that you are having problems of some sort that prevent you from doing this (and almost everyone does), go back to the earlier steps and work your way up to step three and beyond again. For example, a common (if not universal) problem is that the balls start going more and more forward, so that within a few throws you are running forward to catch the next throws. The reason this happens is that as soon as one hand throws a ball slightly forward, the *other* hand has to reach forward to catch it. But that hand is also in the process of throwing a ball, so that, in reaching for the ball it has to catch, it inadvertently throws the next ball even *more* forward. This process is unstable and you end up racing across the room in trying to continue the pattern. The cure is to go back to step two and make sure that the balls don't go forward, sometimes even by forcing them to come back toward your body a bit. Make sure you try this with either hand throwing first. Another helpful hint: Think of throwing the balls more "up" than "across." It they are thrown with too much horizontal motion, the pattern gets so wide that it becomes very hard to even watch the balls, much less control them! In any case, you will find at some point that the three-ball cascade pattern will "click" and you will then wonder why it took you any time at all to learn it.

Of course, this is the most fundamental three-ball pattern. There are literally hundreds of other patterns that can be mastered with three balls, and if you start adding more balls the possibilities multiply exponentially. An excellent book for taking the next steps is *The Art of Juggling* by Ken Benge.[6] A good test to see how well you understand the process of learning this pattern is to teach this process to someone else. In this way, the art of juggling is passed on from one generation of jugglers to the next, in much the same way that magical knowledge has been communicated down through the centuries.

Chapter 10

STARS OF MATHEMATICAL MAGIC
(AND SOME OF THE BEST TRICKS IN THE BOOK)

People have been inventing self-working magic tricks based on simple mathematical ideas for at least a thousand years. In the past hundred years, a revolution has taken place. This comes from the emergence of serious hobbyists. They support magic dealers (many sizable towns have one), hundreds of magic clubs, and about a hundred yearly conventions. Out of this comes progress. There is a constant call for something new. Old tricks are varied, improved, and classified. They are recorded in journals. There are quarterlies, monthlies, and even a weekly journal, *Abracadabra*, which ran for over fifty years. The large magic journals have circulations of five thousand or so. There are many books (and, nowadays, videos), in fact, *hundreds* of magic books are published each year. Finally, there are e-books, blogs, and magic Web sites of every variety—YouTube has thousands of magic snippets. It is a very active whirlpool.

Of course, most of this is chaff: Simple variations, tiny steps forward (and sometimes backwards). Every once in a while, a brilliant new idea surfaces, a good effect whose method is as amazing as the trick itself. There are a handful of inventors who are repeatedly brilliant—we have chosen to discuss seven of them.

Our seven are a varied lot. There is a chicken farmer from Petaluma, a rural free-delivery mailman, a computer wizard (or two), and a priest.

Some led normal lives (a transit engineer) and some were weird (a hobo who lived out of Dumpsters) or even institutionalized. All have created brilliant tricks with deeply original roots, tricks that live on.

Our stars invent mathematical tricks. To put things in perspective, it is useful to ask why the two greatest twentieth-century inventors of card tricks aren't on our list. These two are Dai Vernon and Edward Marlo. Both dedicated long lives to developing esoteric magic. Both were amazingly talented sleight-of-hand performers who could carry out secret maneuvers "in such a manner that the most critical observer would not even suspect, let alone detect, the action." In addition to esoterics, both Marlo and Vernon invented wonderful, performable magic. If you see a street performer do the linking rings or cups and balls, it's likely to be essentially Vernon's routine, move for move. If you see a close-up worker doing card tricks on television, it's likely you'll see some of Marlo's inventions.

Both Marlo and Vernon had "tin ears" when it came to the mathematical end of the world. Vernon told us he was awful at mathematics in high school. In desperation, he tried memorizing the sine and cosine tables! This is a crazy, sure road to failure. Vernon was captain of his schools's hockey team. On graduation, his teammates gave him a ring inscribed with $sin^2\theta + cos^2\theta = 1$ to poke fun at his mathematical failings. He just didn't have a feel for math. His self-working tricks, when they lean on math, are some of his most pedestrian efforts.

Ed Marlo was a machine shop foreman who developed many aspects of modern card magic. He took extensive notes as he worked, writing his tricks on the back of IBM punch cards. We have thousands of these cards as part of Marlo's correspondence.

This correspondence is a remarkable record of card magic in the second half of the twentieth century. Everyone seriously interested in card magic wrote to Marlo and he answered back, usually keeping carbon copies of his answers. We started writing to Marlo at age thirteen—he answered, often with long letters, sometimes with postcards. One of his letters began, "Four years ago 5/2/71 I sent you a postcard with. . . ." A man who answers thirteen-year-old youngsters is saintly. A man who keeps copies of their postcards is a little bit crazy. Marlo too had a "tin ear" for the mathematical end of card magic. He could exploit others' discoveries in brilliant ways—his methods for using perfect shuffles and stay-stack are still in our active repertoire. As far as we know, he never discovered any interesting new mathematical principle.

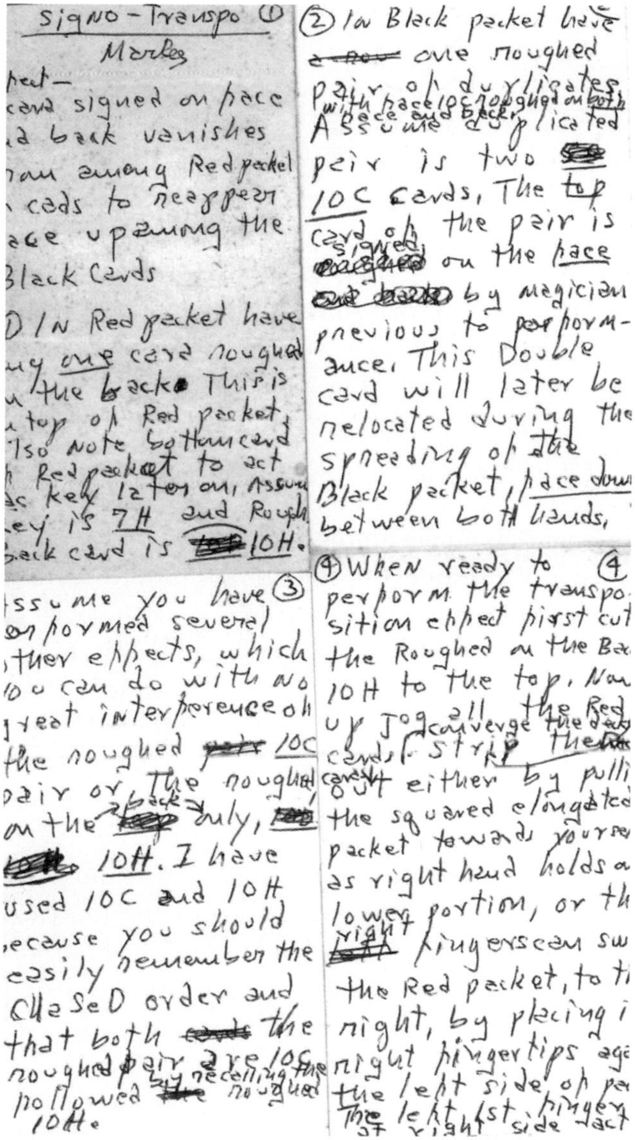

Figure 1. Marlo's notes describing Signo-Transpo

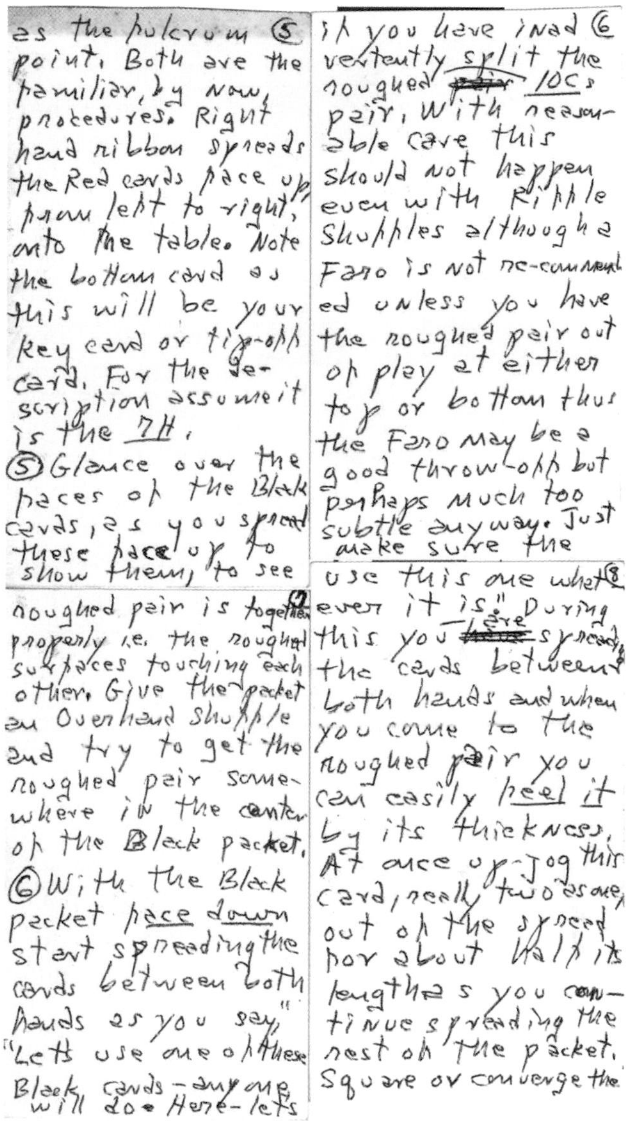

Figure 2. More of Marlo's notes

So, good magic and good mathematical magic are different things. The stars we discuss below created great mathematical magic. There aren't many of them, and they seem to be dying off (and we haven't been feeling so well ourselves, lately!). We hope our focus starts a fire someplace.

ALEX ELMSLEY

Figure 3. Alex Elmsley and aspiring magician, circa 1975

Alex Elmsley was a soft-spoken, kind man with a twinkle in his eye. We first met him in 1958 on a quiet weekday at Lou Tannen's magic store in New York City.[1] Alex had taken a year off after university and the army. He was touring the United States, giving magic lectures, attending magic conventions, seeking to meet the legendary American sleight-of-hand performers. He had invented a simple new sleight called the Ghost Count (now called the Elmsley count), which had taken the magic world by storm. It is one of the few really new techniques to come along and has become a mainstay of modern sleight of hand.

We had our own way of doing the Ghost Count that was less studied and more natural. A long session ensued, and we began swapping stories and tricks. Alex was then twenty-nine, we were thirteen. It is a distinctive feature of magic that an accomplished star can open up to a youngster, with both having a wonderful time. In addition to sleight of hand, Alex loved mathematical magic. He had mastered the perfect shuffle and invented a host of magical applications. Alex taught us the shuffle and some of its basic principles. This was our introduction to binary numbers. On the subway ride home, we wrote out the first eight numbers:

$$000 \quad 001 \quad 010 \quad 011 \quad 100 \quad 101 \quad 110 \quad 111.$$

Alex had explained that if a perfect out-shuffle was carried out for a zero, and a perfect in-shuffle was carried out for a one, the top card could be brought to any position. Alex asked us to think if there was a way to shuffle to bring a card at any position x to any other position y. We have been thinking about it for close to fifty years now. A brief description is in chapter 6. Shuffling cards on a moving train is not the easiest way to start. When people ask how long it takes to do reliable perfect shuffles, we think back to Alex. It might be a few hundred hours, certainly thousands of repetitions, but when you're a thirteen-year-old, the time just flies.

We went back to Tannen's day after day, cutting school to discover some higher truths. Tannen's salesmen, Jimmy Herpick, Frank Garcia, and Lou himself, knew they weren't making any money from us. They loved magic, each in his own way, and let us be.

We next met Alex ten years later in London in 1969. He was teaching advanced programming for one of the big English computer companies. He gave us a tour of the workplace that had a display of old calculators and punched-card readers. We stopped at one of these and he said, "I wonder if they still work? Let's try." He picked up a pile of about fifty or so punched cards and said, "I see, these have names and phrases typed on them. Can you shuffle these?" He handed us the deck of punched cards. We shuffled them thoroughly. "Let's see if the card reader will take them." He dropped them into a table-sized gadget and pushed a button. The gadget chattered along and "ate" the punched cards, spitting them out into a tray. "It seems to work," he said. "Let's try this—cut the cards a few times. Look at the top card. Does it have something written on it that makes any sense?" It said "God save me," a seeming plea from an unfortunate programmer. "Okay, put that card in the middle of the deck, give them a riffle shuffle, a cut, another shuffle and a few more cuts, and drop them into the card reader." He pushed the same button and the aging machine again ate the cards. At the same time, a television screen in the room lit up with a message:

Attention, attention, subliminal message detected—stay tuned.

The screen went blank and, in a few seconds, the message

God save me

appeared. We looked over at Alex—the twinkle was more of a smile than usual. "Care to see it again?" he asked. Of course, it was a setup. This museum was part of Alex's digs and he had prepared several of the gadgets to perform magic tricks. This particular trick is Charles Jordan's riffle shuffle trick (see chapter 5). The idea of doing it with punched cards is brilliant. They can be thoroughly shuffled to start. During the first read through ("Let's see if this works" indeed!) the computer reads the order. Repeated cuts and up to three riffle shuffles don't destroy enough of the cards' order and the selected card can be located. A nice feature—on the second read through, the computer records the current arrangement. Not only is it set to reveal the chosen card but it is also set for an instant repetition. Alex's smile was a knowing one.

Alex had thought of lots of computer tricks. A simple one is to ask someone to name a card out loud and then have him or her type "What is the name of my card?" into a computer. After some byplay, the computer displays the card. The method here is that you have fifty-two different ways of asking the question (with or without a question mark, with the phrase "card please," instead of "my card," etc.). He marketed an amazing extension of this that has to be seen to be believed. When performing for a group of programmers, he has been known to have a secret assistant type in some extra information in another room. An equally devious ploy is to have the computer perform a "psychological force" on the spectator. Such forces are usually carried out by a forceful performer who influences the spectator to think of a predetermined card while seemingly offering a free choice (Alex used Vernon's Five Card Mental Force, for those who know what the words mean). This combination of mathematical, psychological, and external methods can create a truly spooky, unsettling effect.

We saw Alex about once a year on visits to London. For most of his life he lived at 6 Smith Terrace, the family home. Alex stayed on, looking after his mother. We caught a glimpse of her when we tiptoed into the living room after midnight. "Alex, who is your friend?" she asked. "Oh, this is Persi Diaconis. He was kind to me on my trip to America." Alex's mother smiled politely and left us to our own devices.

Alex's mother died at age 103. In the ensuing mess the house had to be sold, Alex moved, and then was truly alone for the first time. This was clearly a tough period. One aspect came out during a trick. We were performing the trick based on the *I Ching* described in chapter 8. At one point, the spectator (Alex) had to ask the book a question. He asked, "Should I get a dog?" in a way that made it clear it was a deadly serious question for him. We don't remember what the *Book of Changes* said but, on further inquiry, Alex said he had lost his keys and been locked out two nights before. Not daring to roust one of his new neighbors at 11:00 p.m., he slept in the foyer, calling a locksmith the next morning. The image of the greatest inventor of our kind of magic seemingly without a friend to call really hit home.

There were wonderful up times too. Once, we met Alex with our friend Ricky Jay in the cafeteria of the British Museum. After doing card tricks for an hour or so, we settled into conversation. Alex asked if we still had his letters to Dai Vernon (the ones we saved from the fire, but that is another story). We certainly did. Alex reached into his pocket and pulled out a thick packet. "I thought you should have the rest of the conversation." With that, he handed over Vernon's letters to him, a treasure trove on both sides.

Alex went up and down about magic. When up, he bubbled over with new ideas and enthusiasm for new tricks he had seen. When down, we found other things to talk about. Once he asked if irrational numbers (such as $\sqrt{2}$) were really needed. He had no doubt they existed but had become intrigued by the following question: No quantities in the real world can be measured with infinite precision. Take a bunch of points in the plane and put a tiny, tiny circle about each of them. The circles represent measurement error; we only know each point up to some tiny error. Could it be, he asked, that we can always find a point within each circle so that all the distances involve only rational numbers? This seemingly innocent question turns out to be beyond modern mathematics to answer. Indeed, it is unknown if one can find *eight* points with no three on a line, and no four on a circle, so that all the distances involved are rational numbers.[2] On the other hand, it cannot be ruled out that there could exist a *dense* subset of points in the plane so that all the distances between them are all rational.[3]

We have not discussed many of Alex's tricks above. Most of these are lovingly described by Stephen Minch in the two-volume *Collected Works of Alex Elmsley*.[4] These volumes contain personal history, wonderful nonmathematical tricks, and dozens of new mathematical tricks. Most of these have yet to be abstracted. One of the joys of the modern era is that great magicians are being videotaped. A four-volume DVD set of Alex performing is also available.

We close this section on Alex by describing just one of his discoveries, what he called Penelope's Principle. To set the stage, we recall a spirited discussion between your authors, a great sleight-of-hand expert (Charlie Miller), and the publisher of a magic magazine. The publisher asked: "What's the big deal about the perfect shuffle? The only way it is actually used is as a kind of glorified false shuffle. Then, its kind of studied and more classical methods are better." We responded by explaining Penelope's Principle. This shut the publisher up and had Charlie Miller wanting to try it out. We will *not* explain it in its perfect version.[5] Instead, we explain a less sleight-intensive version discovered by Dai Vernon. This was in Vernon's correspondence with Elmsley (October 25, 1955). It is explained here for the first time.

In essence, the performer writes down a prediction, with a deck placed face-down before the trick begins. The spectator cuts off more than half the deck, leaving the small packet on the table and handing the larger packet to the performer. The larger packet is milk shuffled onto the table, a pair at a time (see chapter 6). Well before this is finished, the spectator is invited to say stop at any time. The remaining cards are dropped back on top of the larger packet. The prediction is now turned over. Suppose it says "jack of hearts." The performer removes cards simultaneously, one at a time, from the tops of both the large and small packets until the small packet is exhausted. The final card dealt from the larger packet is turned up. It is the predicted jack of hearts.

The trick works itself. All that is required is for you to spot the card originally twenty-sixth from the top and write this down as a prediction. In working, roughly estimate how many cards the spectator cut off. As you milk through the larger packet, offer to stop as directed and drop the remainder on top only when you have fewer than the estimated number remaining.

Here is a first illustration of how this may be built up. Have any even number of cards set in correspondence, e.g., cards half a pack away match in color and value. Thus, with a ten-card pack, the cards might be arranged as AS, 5D, QC, 4H, 7H, AC, 5H, QS, 4D, 7D. Have the cards repeatedly cut by the spectator, who then hands you a random amount, more than half. These are milk shuffled as described above and the big packet is tabled. The spectator's small packet and the tabled packet are turned up simultaneously. The very last cards match. Any of the perfect shuffle tricks involving Penelope's Principle can be adapted to the much easier to execute milk shuffle.

BOB NEALE

Bob Neale spent much of his life as a professor of psychiatry and religion at New York's Union Theological Seminary. He is also an ordained minister who worked for several years in England at a hospital for terminal patients. If you (or a friend) are trying to think about death and dying, we recommend *The Art of Dying*, which Bob wrote as a workbook for his patients. (Incidentally, for thoughts about death, written thirty years ago by one of your authors, you can check out *Death and the Creative Life: Conversations with Prominent Artists and Scientists*, by Lisa Marburg Goodman.)[6]

Turning a corner, Bob is one of the world's great creators of paper folding and origami. We don't know about you, but our reaction when someone says they do origami as a hobby is, "Oh no, we're about to be shown little crumpled lumps of paper with 'legs' sticking out, followed by 'It's a dog.'" Bob's creations are different, however; a few deft folds

Figure 4. Bob Neale

and an abstract form that is clearly a "nun" shows up. Other folds are impossible objects, often made from a single sheet of paper with no cuts or gluing allowed. Origami is an art form in Japan and Bob is one of the very few Westerners whose paper folds are taken seriously there. We are *not* paper-folding aficionados, but recently Bob showed up at one of our haunts to visit a small puzzle party. Word somehow got out and folders from hundreds of miles around showed up. They are mostly a gentle, nondemonstrative sort but even an outsider was inspired by the respect Bob commands from this audience.

Turning the next corner, Bob has a professional knowledge of anthropology, folklore, philosophy, psychology, and psychiatry. He sees themes and depths in magic and can bring these alive to workaday pros and serious amateurs. He thinks and writes about what makes magic tick, what makes good magic move people, and the tensions between a magical experience and being fooled.

Bob's theatrical bent illuminates his teaching. One of these interactions is worth recording. Two groups of graduate students made class presentations for their final projects. They took place in Bob's spacious apartment on New York City's Riverside Drive. The theme of both projects was EVIL and the first group wasted no time getting into it. They had built a small but real guillotine and had captured a stray cat. They argued that the life of an alley cat was not much and unless the rest of the group had arguments to the contrary, they were going to actually behead the cat. They skillfully deflected the few arguments offered and then *boom*—they did it. The shock of this happening for real touched all in the room. In their debriefing, they owned up; the cat had terrible cancer and was to have been euthanized the next morning. Heavy sedation ensured it had no pain. They wanted to bring the awful feeling of evil alive in a way that transcended conversation. They succeeded.

The second group of students proceeded as follows. They broke into small groups with the audience members and spread out into several rooms for quiet, personal conversations. After about half an hour of this, they started to reassemble the groups. The conversations stopped and the students began to report personal, embarrassing things they had learned from the audience during their "friendly bonding." People's failings, strong negative statements privately uttered about others in the group, and worse were brought out into the open. It was mean and maddening. It turns out that their aspect of evil was "betrayal." They brought out its evil nature in a way that no one present will forget.

If Bob can do that in an ethics and religion class, just imagine what he can do in the freer space of a magical performance. It's hard to imagine but fortunately he has written a lot of it down.

Before we launch into tricks, here is a story that contains a wonderful piece of performing material. One of us was stuck with Bob in an airport delay. A six-year-old girl, looking very sad, was sitting near us. Out of the blue, Bob turned to her and said, "Would you say hello to my friend?" With this he took out a sheet of 8½" by 11" paper. The girl was shy but curious too. She kept looking. Bob folded the sheet a few times, made a tiny tear (heaven forbid), and drew some circles. He folded a few more times and bingo, a cute little beak puppet appeared. "Hello," the puppet said. The little girl said "Hello" and her face lit up into a big smile. It was not only the girl's smile that remains in mind twenty years later. Seeing Bob fold a sheet of paper is like seeing an expert card handler shuffle a deck of cards. It's simply beautiful. All these years later, we've asked Bob to teach this magical moment. We asked him not to hate us for being so fond of this trivial fold that breaks all the rules. He knows it's magical too. A sequence of photos teaching this fold is shown in figure 5. A clear description of the classical "snapper" beak puppet is in Robert Harbin's *Paper Magic*.[7] It's not hard, but practice makes the folds fold smoothly. Bob allows that his other "performance fold" is his version of the jumping frog, shown in figures 6 and 7 (see "Bullfrog" in his book *Paper Money Folding*).[8]

Bob's magic often uses topology, the mathematics of deformed shapes where a coffee cup and a doughnut are considered indistinguishable. He has ways of tying a knot in a string without letting go of the ends and making short squat tubes of paper that turn into long skinny tubes with a shake of your hand. He also has a wonderful presentation piece in which a spectator, behind bars, gets out of jail while holding onto the bars at all times. These tricks are expertly described in Bob's books. His most recent, *This Is Not a Book*, contains many mathematical gems that also make for good performances, along with pointers to the rest of Bob's publications.[9] We describe his Rock, Paper, Scissors trick in the section on Bob Hummer, later in this chapter.

Turning the corner to magic, we offer a topological trick of our own devising that Bob likes and contributed to. We have kept it secret for fifty years, performing it only for "real people." It is our version of the endless chain, a gambling swindle still in use on the streets of large cities.

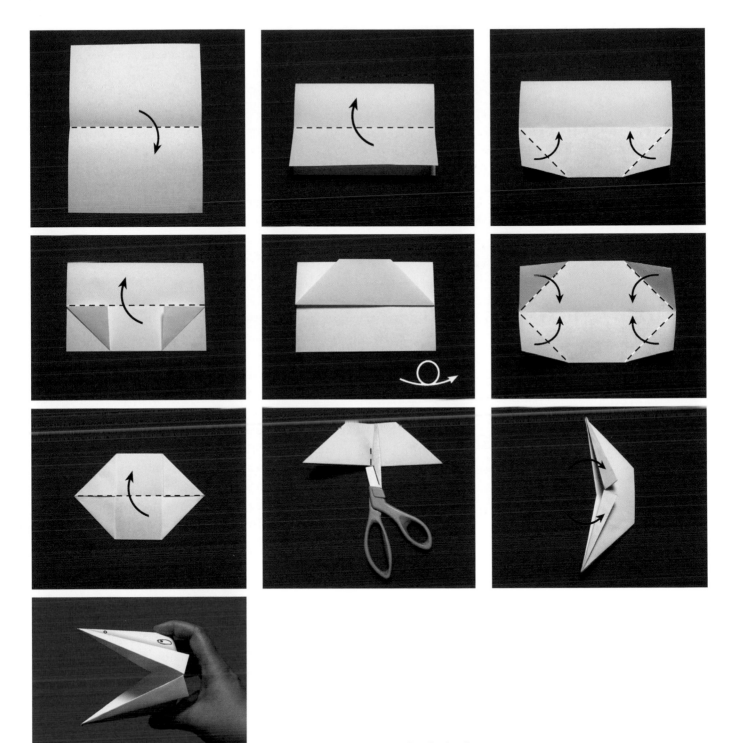

Figure 5. Folding sequence for the beak puppet

Figure 6. Resting frog

Figure 7. Frog about to jump

Figure 8. Bob teaches Kate Graham to make a flapping bird

ÍNSIDE — OUTSIDE

The performer shows a closed chain loop and opens it up into a circle on the table. A participating spectator is asked to put a finger inside the loop, touching the table (figure 9).

The performer picks up the chain and moves it about, explaining, "No matter how this is wiggled around, it stays hooked on your finger (unless it's run up your arm and over your head). Similarly, if your finger is outside the loop" [the spectator complies], "fiddling with the chain won't get it hooked on your finger."

The performer continues: "Suppose we drop the chain into a random mess on the table. If you put your finger into the mess, it's hard to say if it's inside or outside." The spectator again complies (figure 10). "What do you think?" the performer asks. The spectator guesses "inside" or "outside" and the performer pulls the chain until it either is caught on the spectator's finger or pulls free.

"Let's make a little game out of this," the performer patters. "Look," the chain is placed into a loop and the performer touches the table at points A, then B, then C, saying, "outside, inside, outside. Right?" (See figure 11.)

Now, the end of the chain is pushed into the circle, as in figure 12. The performer touches A, then B, C, and D, saying, "outside, inside, outside, inside." Next, the loop at D is picked up and placed onto the chain between A and C (figures 13 and 14).

The performer puts a left finger at X and a right finger at Y (figures 15 and 16), pulling apart, keeping the loops at X and Y open (figures

Figure 9. Finger in a simple chain loop

Figure 10. Finger in a tangled chain loop

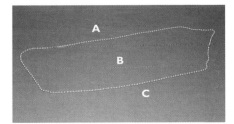

Figure 11. Chain on the table

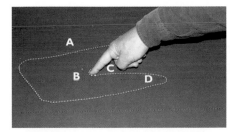

Figure 12. Pushing end into the circle

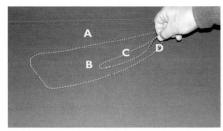

Figure 13. Picking up loop at D

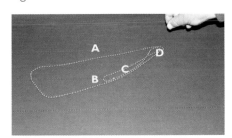

Figure14. Placing it on the loop

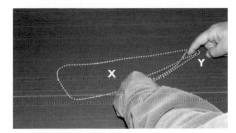

Figure 15. Putting fingers in loops

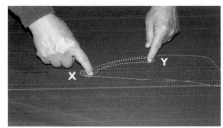

Figure 16. Different view

Figure 17. Pulling apart

Figure 18. Completed pulling apart

Figure 19. Squeezing together

Figure 20. Final configuration

17 and 18). The top and bottom of the chain are pushed together as in figures 19 and 20.

The performer continues: "One loop is inside; if you put your finger there, you win. One loop is outside; if you put your finger there, you lose." The spectator is asked to choose, puts a finger down, and the chain is pulled until it pulls free or is caught. Almost always, the spectator puts a finger at Y and wins. "Hmmm, you've played this before?" the performer asks. "Let's try it again." The whole sequence is repeated, perhaps pushing in from the left this time. The spectator will probably guess correctly and the performer acts chagrined.

"Okay, let's play it for real now." The performer makes a tricky throw, winding up with the basic configuration (figure 19). Now it is not obvious which side wins. In fact, this third throw is made so that *both* sides win. We explain this subterfuge in the following section. It is the creation of the great American magician Stewart Judah. The performer looks cunningly at the spectator, "One side wins, one side loses, which do you pick?" The spectator chooses and the performer (hamming it up a bit) asks if the spectator wants to switch. Whichever choice is finally made, when the chain is pulled, the spectator wins by having her finger caught. The performer acts defeated, "Are you sure you haven't played this before?"

Now, changing the tone slightly, the performer offers to play for one penny. The chain is thrown, the spectator chooses, and the chain pulls free (the spectator loses). A practice throw is made, and the spectator wins. Another throw for a penny—the spectator loses. This can be continued. The point is, the performer has several ways of throwing the chain (explained below). One ensures that both sides win, another ensures that both sides lose. There are similar-looking throws where left wins and right loses (or vice versa).

After a bit of this the performer patters, "These throws are complicated. Let's go back to basics." The chain is made into a circle and the same steps (and patter) as in figures 10 through 19 are carried out. By this time, the spectator is getting used to these configurations and it is clear which side wins. The moves are made slowly and deliberately. The performer offers to bet double or nothing. We have had spectators—sure they were going to win—offer to bet a lot more. When the spectator chooses, the chain is pulled free and the performer wins. Indeed, there is a final crafty twist that ensures that both sides lose (after all, it is double or nothing!). For details, see the following section.

The sequence above can be presented as a gambling game, as a lesson in how to protect yourself on the street, or as a story. The performer can involve several spectators, with one as a shill who always wins. Some performers like to challenge the spectator while others try to avoid making anyone look foolish. The endless chain is a compatriot of three-card monte and the three-shell game. All are widely used hustles even today. There is a lot of magical literature on how to entertainingly perform with various montes. Much of it can be adapted to the endless chain.

It is "obvious" that a closed circle, placed on the table and wiggled about, has an inside and an outside. In the variants discussed above, the chain is a three-dimensional object that goes above and below itself on the table. Consider the simpler case where the chain is first in a simple circle and poked about, without crossing, in an arbitrarily complex fashion. Clearly, it still has an inside and an outside. It turns out that this is really hard to prove rigorously. This is the celebrated Jordan Curve Theorem. It is usually carefully proved only in a graduate course in topology. The most accessible proof we know is in Carsten Thomassen's wonderful article "The Jordan-Schönflies Theorem and the Classification of Surfaces."[10] A recent trend in mathematics is to try to give proofs that are completely free of heuristic geometric reasoning. Every step of these proofs is carefully checked by computers. The Jordan Curve Theorem was finally given a computer-based proof in 2005 by Thomas Hales. The history of formal proofs in mathematics is engagingly told in the December 2008 issue of the *Notices of the American Mathematical Society*.

THREE FINAL SECRETS

We now explain several throws of the chain. Some are "fair" in that one side wins and one side loses. A "super-fair" throw has both sides winning. The "cheating" throw has both sides losing.

A FAIR THROW. The vanilla flavor fair throw begins with a circle, the performer loops it around, lays it on top of itself, and pulls it into a final configuration (see figures 21 through 28).

We do this by placing the right hand under the original circle at the right, carrying the right-hand side around to the left, placing the original right on top, and then straightening this out to the final configuration as shown. With this throw, the right side loses and the left side wins. If you put the left part of the original circle on top of the right

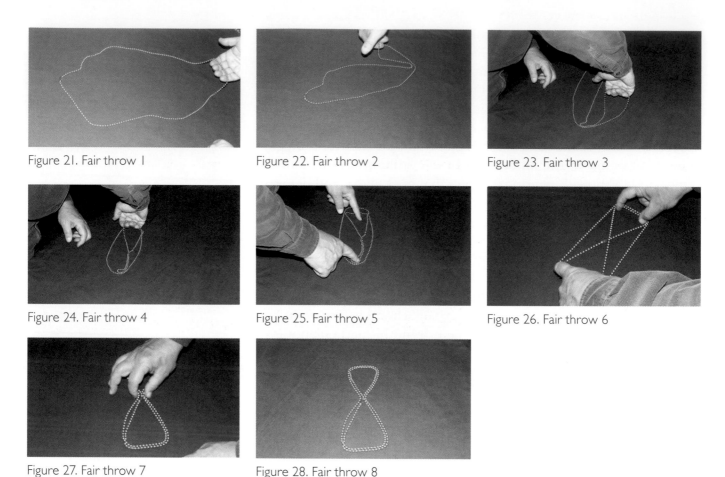

Figure 21. Fair throw 1

Figure 22. Fair throw 2

Figure 23. Fair throw 3

Figure 24. Fair throw 4

Figure 25. Fair throw 5

Figure 26. Fair throw 6

Figure 27. Fair throw 7

Figure 28. Fair throw 8

in a kind of mirror image, the left side loses and the right side wins. There is no skill required to perform these throws. A small amount of practice will make them flow.

A CHEATING THROW. This is almost indistinguishable from the fair throw and results in both sides losing. Beginning with a single circle, use your right hand to lift up the right-hand piece (right hand palm-up) and place it on top of the left part, turning palm-down as you do so (see figures 29 through 36).

A SUPERFAIR THROW. This was shown to us fifty years ago by the Cincinnati magician Stewart Judah. One of the original *Greater Magic* Ten-Card stars, Judah was soft-spoken and slow working. Most of his

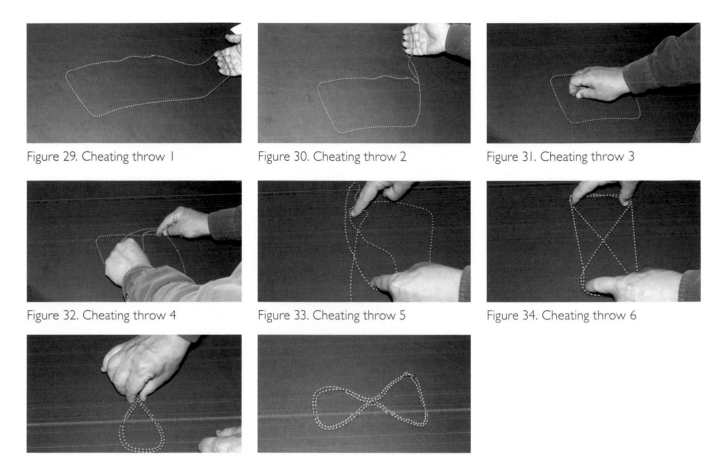

Figure 29. Cheating throw 1

Figure 30. Cheating throw 2

Figure 31. Cheating throw 3

Figure 32. Cheating throw 4

Figure 33. Cheating throw 5

Figure 34. Cheating throw 6

Figure 35. Cheating throw 7

Figure 36. Cheating throw 8

tricks did not depend on advanced sleight of hand, just diabolical handling. As far as we know, this maneuver has not been previously described. However, Nick Trost, a student of Judah's, recorded many of Judah's wonderful inventions. Trost's book describes many of Judah's gambling demonstrations, including a handling of the belt loop, an early version of the endless chain.[11] However, Judah's endless chain moves are not described. Instead, an interesting endless chain originated by Dennis Flynn is described. This has a very different set of throws, done in the air, for achieving the same four final outcomes. Without further ado, Judah's throw is illustrated in figures 37–42.

Begin with the chain in a circle, flat on the table. Place both hands, palms up, under the part of the chain farthest from you (figure 37). Curl both hands inward, so that your fingers point in towards you,

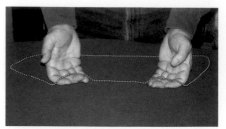

Figure 37. Judah 1

Figure 38. Judah 2

Figure 39. Judah 3

Figure 40. Judah 4

Figure 41. Judah 5

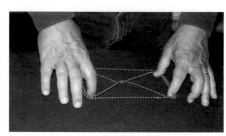

Figure 42. Judah 6

making two small loops (figure 38). Cross your hands, keep both palms upward, and lay the two loops at opposite ends of the chain (figure 39). Drop the loops onto the table, removing your hands (figure 40). Now straighten the chain into the basic figure-eight position as usual (figures 41 and 42). If done properly, both sides are winning for the spectator.

Synopsis. There is a simple mathematical underpinning to the three throws just described. In each, the right side of the original loop is laid on top of the left. In the first variant, the right side is kept parallel to the left at all times. This results in the right side losing and the left side winning. In the second variant, a single twist is made in the right side before laying it on the left. This results in both sides losing. In the third variant, a double twist is made in the right side before laying it on the left. This results in both sides winning. With a bit of thought and practice, these extra twists can be done subtly so that they are not noticed by the audience.

Denouement. In the last sequence of the performance, the performer returns to the simple beginning sequence shown in figures 11 through 14. The only difference comes at the end. Instead of laying the bottom loop on the top, keeping it simply parallel, turn over the bottom loop (see figure 32) as you lay it on top. This is done in a completely open fashion, and the final result has both sides losing. If each of the top and bottom loops is given a twist as you lay them on top of each other, the result in that both sides win for the spectator.

With all this control, you can let the spectator decide if being caught or free results in winning or losing. This can even be varied from throw to throw. In performance, this much freedom is confusing. It is best to stick to the classical "catching wins." We will not detail further performance tips but many spectators think that the way the chain is pulled away affects things. You can let them pull the chain away. We sometimes let the spectator put one finger on each side, pull the configuration taut, and, after study, let her lift one of her fingers.

A Dai Vernon Trick. We cannot resist adding a trick to the discussion. The trick may well be standard, but it was a favorite of Dai Vernon, the greatest exponent of pure sleight of hand of the twentieth century. Vernon had collected fifty impromptu tricks done using a piece of string. We don't remember the other forty-nine very clearly but one has stayed with us for over fifty years. People find it surprising.

Take a loop of string knotted into a circle (or the endless chain of the previous demonstration). Have someone hold up a finger and loop the chain once around it. You hold the other end of the chain in your left hand. There should be some tension that holds the chain fairly taut (figure 43).

The performer's right hand comes over the chain from above, palm-down. The right-hand second finger contacts the left part of the chain at point X (see figure 44) and moves it right and over the right-hand part of the loop (figures 45 and 46). Now, the right hand turns palm-up, twisting its piece of the chain under the other side (figure 47). The right hand's index finger enters the loop adjacent to the spectator's finger at point Y (figures 48 and 49).

Now, with the chain entangled between its fingers, the right hand turns back palm-down (figure 50). The tension between the left hand and the spectator's finger keeps the configuration from slipping. Place the right hand's second finger on top, so that it is touching the spectator's finger (figures 51 through 54).

Figure 43. Starting position

Figure 44. Close-up of starting position

Figure 45. Finger inserted

Figure 46. Starting to move finger

Figure 47. View from below

Figure 48. Approaching the other hand

Figure 49. Inserting the index finger

Figure 50. Releasing the third finger

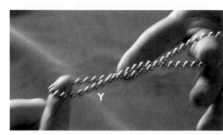

Figure 51. Connecting the fingers

Figure 52. Releasing index finger

Figure 53. Index finger is released

Figure 54. Chain is released

This can be presented by saying something like, "This loop is constrained at both ends. What I do with the middle won't change that. Watch . . . !" People find it quite magical.

HENRY CHRIST

Henry J. Christ was an inner-inner-circle magician. He was close to the great sleight-of-hand performers of the twentieth century—masters like Leipzig, Cardini, Annemann, Findley, Vernon—who created modern magic. Henry knew them, caroused with them, and fooled them. All of these experts coveted Henry's secrets and showed him their best in exchange.

Born in 1903 in Brooklyn, New York, Henry started doing magic very early. His hands were small at seven years old and his parents bought him a miniature deck of Little Duke playing cards to practice with. He invented and performed magic for friends. An early magical bill for performing college students shows Henry demonstrating card flourishes.

Henry married Evelyn Pilliner, had three kids (Richard, Michael, and Robert), and took a job as an engineer with New York's Transit Authority. He held this job for more than forty years (1924–1968), then retired to practice the guitar and his beloved magic. The great sleight-of-hand performer Sam Horowitz (also known as Mohammed Bey) was at Henry's retirement dinner. Figure 55 shows how they looked in 1968.

Henry kept magic in place as a serious hobby. He invented all sorts of tricks—for example, the color-changing deck. Here, a card is selected and shuffled back into the deck. All of a sudden, the backs of all the cards change from red to blue and the spectator's selection is the only red-backed card left in the pack. Henry showed this trick to the great stage performer Nate Leipzig, who popularized it. The trick is widely performed today.

The first avant-garde magic magazine was Theo Annemann's *Jinx*. He pried the best secrets out of the best performers and preserved these in a four- to six-page biweekly format. Annemann was a real expert at pseudo–mind reading and his treatise *Practical Mental Effects* is a bible for this work. Henry and Annemann became best friends when Annemann moved to New York. Annemann had shaky emotional health and a serious alcohol problem. The *Jinx*'s coming out on time and much of the magic it contained depended on Henry.

Fligure 55. Henry Christ (left) and Sam Horowitz, circa 1968

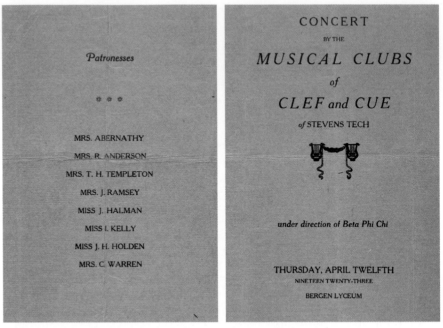

Figure 56. Outside of program with Henry Christ performace

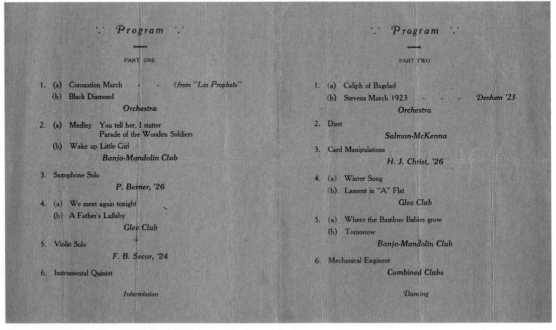

Figure 57. Inside of program with Henry Christ performace

Some of Henry's tricks are described in the pages of *Jinx*. One celebrated trick, Dead Man's Hand, is built around the death of Wild Bill Hickock, while he was playing poker. He was shot in the back while holding the "dead man's hand" of aces and eights. Henry's trick weaves around Hickock's story. At the end, when the aces and eights are shown, a shot rings out—a really loud bang—made by a cap pistol under the table. This gives the trick a dramatic ending. Henry's recollections of this trick (in his own words) are shown in figure 58.

Notes on "Dead Man's Hand" (by HJC)
published in *Jinx* #33, June 1937

A few days after the publication of *Jinx* #33,
I was at Gen Grant's magic shop on 42nd Street.
As I walked in, David Bamberg was making a purchase
at the counter. After he completed his purchase,
he left the counter. He had a copy of
Jinx #33 in his hand and a paper bag. He
recognized me and said, "Are you the Henry
Christ who dreamed up this 'Dead Man's Hand'?"
I said I was. He said "This is the greatest
dramatic card effect I ever heard of! I
just bought 24 cap shooting gimics [*sic*]. This
is one trick I'm going to do!" This was
indeed a complement [*sic*] coming from a performer
of his background and experience!

About a month later, I met a friend
of mine, Hal Haber, a devout amateur
magician. He was a successful consulting engineer, had
degrees in Chemical, Mechanical, Civil and
Electrical Engineering and was a Licensed
Professional Engineer. He could make a
perfect rising card mechanism out of any
one dollar Ingersoll watch! He said to
me, "I did your 'Dead Man's Hand' the other
night at a Newspapermen's Convention. They
are a hard-boiled audience. But the 'Dead Man's
Hand' knocked them for a loop. They said it was
the greatest card trick they ever saw!"

Notes on "Dead Man's Hand" (by HJC)
published in Jinx #33, June 1937.

A few days after the publication of Jinx #33,
I was at Gen Grant's magic shop on 42nd Street.
As I walked in, David Bamberg was making a purchase
at the counter. After he completed his purchase,
he left the counter, ~~someone~~ He had a copy of
Jinx #33 in his hand and a paper bag. He
recognised me and said, "Are you the Henry
Christ who dreamed up this "Dead Man's Hand?"
I said I was. He said "This is the greatest
dramatic card effect I ever heard of! I
just bought 24 cap shooting gimics, This
is one trick I going to do!" This was
indeed a complement coming from a performer
of his background and experience!
    About a month later, I met a friend
of mine, Hal Haber, ~~a~~ a devout amateur
magician. He was a consulting engineer, had
degrees in Chemically, Mechanical, Civil and
Electrical Engineering and was a Licensed
Professional Engineer. ~~He~~ He could make a
perfect rising card mechanism of out of any
one dollar Ingersoll watch! He said to
me "I did your "Dead Man's Hand" the other
night at a Newspaperman's Convention. They

are a hard-boiled audience. But the "Dead Man's
Hand" knocked them for a loop. They said it was
the greatest card trick they ever saw!"

Figure 58. Notes from Henry Christ

Henry invented and performed great sleight of hand, mechanical magic, and the cups and balls (with differently colored balls and color-coded cups). He found joy in subtle, self-working card magic. Some of his tricks are recorded in Martin Gardner's *Mathematics, Magic and Mystery*. Henry tired of magicians in the 1950s and restricted himself to seeing only New York's inner circle. He'd had enough tricks stolen and published under someone else's name that he stopped showing tricks to magicians.

We met him in 1960 when Vernon brought us around to the inner circle. Henry showed us a trick called Eliminating the Methods. Four aces are placed separately on the table, with a few cards placed on each ace. At the end, the four aces assemble in one pile. The trick is designed to fool magicians. As the cards are dealt and handled, the standard methods of doing the four-ace assembly are described and eliminated—"Sometimes performers only pretend to deal the aces on the table. To eliminate this, I will deal them face up in separate spots. Some people use extra aces. To eliminate this, let's deal the extra cards so you can see what they are. . . ." At the end, the aces assembled anyway and none of the experts had a clue as to how it worked.

We became friends with Henry, engaging in a long, detailed correspondence about his magic. He carefully taught us close to one hundred of his tricks. One of the last was the details of Eliminating the Methods. We hope to do these wonderful tricks justice someday, but that is not for this book. We content ourselves with a trick we invented jointly with Henry.

THE ROULETTE SYSTEM

The performer says he has learned an amazing system that guarantees that he will always win at roulette. To demonstrate, a prediction is made and handed to the spectator for safekeeping. The roulette wheel is simulated by a packet of five cards: Two reds, two blacks, and the joker (a house number). "I'll bet on whatever you say, red or black, and I lose when the joker comes up." The spectator decides on (say) red. The performer lays out the system:

> It's kind of a martingale, but with a twist. Whatever I bet, if I win, I'll bet half as much the next time. If I lose, I'll bet half again as much. Look: I'll bet $16 first. If I win, I'll bet $8 next. If I lose, I'll bet $24 (that's $16 plus half of $16).

The spectator shuffles the little deck of five cards and the cards are turned up one at a time. Each time, the performer bets on, say, red, following the rules. A tally is kept and, at the end, no matter how the cards are shuffled, the performer wins exactly $5. When the prediction is opened, it says "I win $5."

Suppose the cards come out R, B, B, J, R (J for joker). Then the tally is:

Bet	Win	Lose
16	16	
8		8
12		12
18		18
27	27	
	43	38

Hence, the difference is $43 − $38 = $5, just as the prediction says.

The trick is self-working and permits many variations. As a check, suppose the cards come out B, R, R, B, J. In this case, the tally is:

Bet	Win	Lose
16		16
24	24	
12	12	
6		6
9		9
	36	31

Again, the difference is $36 − $31 = $5, as predicted. In performance, we sometimes write the prediction on one side of a piece of paper, placed face-down on the table. The tally sheet is written on the other side and filled in as the game progresses. There is no reason to use cards. Five coins or counters—two pennies, two dimes, and a quarter for the joker will work just as well. The spectator can even be allowed to cheat, looking at the cards and deciding which ones to turn. As long as all of the cards are turned once, the final result is forced.

This trick began, as so many do, with a postcard from Martin Gardner. One of his readers had sent in a curious puzzle and he thought a trick might result. We talked it over with Henry and that is how the trick above evolved. Even back then, all of us were interested in "getting to the guts of it." Here is what we discovered.

WHY IT WORKS (AND SOME VARIATIONS)

Suppose we have a deck of N cards with k red cards and $N - k$ black cards. We bet on red to win throughout. In our example, $N = 5$, $k = 2$, $N - k = 3$. (We count the joker—a losing card—as black here.) The betting system is this: Each time, if the last bet is A and we win then we next bet xA. On the other hand, if we lose then we instead bet yA. Here, x and y are positive numbers. (They were $\frac{1}{2}$ and $\frac{3}{2}$ in the example.) We go through the cards one at a time until the end.

Here is the main finding.

If $x + y = 2$ then no matter how the cards are shuffled, the final amount won is the same. This final amount is $\frac{1 - x^k y^{n-k}}{1-x}$ (or "winning factor") times the original-amount bet.

For our original example,

$$\frac{1 - x^k y^{n-k}}{1-x} = 2\left(1 - \frac{27}{32}\right) = \frac{5}{16}.$$

So, if the original-amount bet is \$16, this yields a win of \$5.

After the discovery, we have found a simple way to see the final result. Suppose the original order of the cards is some scattered pattern $RBBRRB.\ldots$ We show that switching a consecutive pair BR to the pair RB yields the same win. Thus, all the reds can be shifted left and the final patterns $\underbrace{BB\ldots B}_{k}\underbrace{RR\ldots R}_{n-k}$ has the same final win. This is now easy to compute and yields the final answer. For example, if the initial arrangement is $BRBBR$, the successive changes to $RBBBR$, $RBBRB$, $RBRBB$, $RRBBB$ all yield the same final win.

Why? Consider a bet of A and successive cards in order RB and BR. The following calculation shows the outcome for the arrangement RB: Initially we win A, then bet and lose xA for a total win of $A - xA = (1 - x)A$. For the arrangement BR, we initially lose A and then bet (and win) yA for a total win of $-A + yA = (y-1)A$. Note that if $x + y = 2$ then $1 - x = y - 1$ so the total wins are the same. From here, the final calculation is easy (it involves summing a geometric series). Also, after either RB or BR, the amount bet is the same, namely, xyA.

The calculations and final formulas have some consequences. Here are two. With five cards, two winning and three losing, using $x = \frac{1}{2}$ and $y = \frac{3}{2}$, gives a final win of $\frac{5}{16}$ times the initial bet. We may ask what value of x gives the best results. It turns out that $x = 0.07777\ldots$ (a little more than $\frac{1}{13}$) gives better results, with a final win of about 1.0377 times the initial bet. For the second consequence, consider a deck with only one winning card and $n-1$ losing cards. It doesn't look good for the bettor. Nevertheless, there is a choice of x (and $y = 2 - x$) such that no matter how the cards are shuffled, the performer wins a fixed amount of the initial bet. When $n = 5$ and $k = 1$, the best x is actually 0, and this guarantees a final win equal to the amount initially bet. What this means is that the performer should keep doubling the bet after each loss ($y = 2$) and, after the unique win (since $k = 1$), should continue the rest of the game by betting 0. When $k = 2$, then for any value of n, the best value of x is slightly more than 0, and this choice guarantees a win of slightly more than 1 times the initial bet. In figure 59 we give a plot (for $k = 2$) showing how the winning factor varies as a function of x, so that when $x = 1/2$ then our winning factor is $\frac{5}{16}$ (as we saw in our

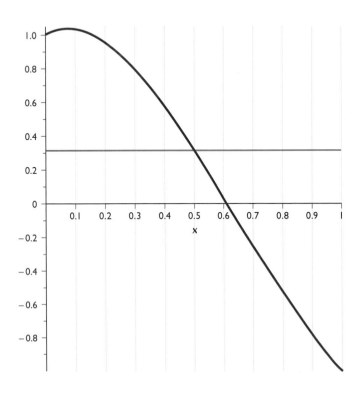

Figure 59. Plot of "winning factor" versus x

example) whereas for $x = 0.07777\ldots$ the winning factor is somewhat larger than 1. This phenomenon holds for any value of $n > 2$.

In fact, when $k = 2$ then, for any value of n, the best value for x (which is usually just slightly larger than 0) always guarantees a winning factor of (slightly) more than 1.

If we take $x = 2$, the performer keeps doubling bets until a win and then stops betting. The formula shows the total amount won will always be the initial bet as long as there is at least one winning card. We caution the reader that all of this depends on our setup of sampling *without* replacement. If the cards are reshuffled every time, there are no gambling systems that beat the house advantage.

There are a number of gambling games in which the final answer is forced, independent of the strategy chosen. In the game of Say Red, a deck of n reds and n blacks is well shuffled. The cards are turned up one at a time. The spectator may say "red" before any turn—even the first turn—and must say red, once, some time. If the next card is red, the spectator wins \$1. If the next card is black, the spectator loses \$1. Since the spectator can say red before any cards are turned up, the game is (at least) fair. It *seems* clear that the spectator can gain an advantage by waiting until there are more reds left than blacks. Cornell University's geometer Robert Connelly's surprising discovery is that it simply doesn't matter. Whatever complex strategy is devised (as long as it doesn't involve looking into the future), the game is exactly fair.

The mathematics involved in setting this up and proving it is the same as the mathematics used to prove that our roulette trick works. The same math can be used to show that gambling systems are impossible. The reader can find out more by tracking down the phrases "martingale" and "optimal stopping."

A similar surprise (i.e., no matter what strategy is used, the final outcome is predetermined) occurs in the world of "chip firing" games, for example.[12] We have yet to see a reasonable, performable trick based on these developments. We are sorry we can't ask Henry.

STEWART JAMES

I hated school, everything about it, and mathematics most of all.

The stars in this chapter have produced strangely brilliant tricks that leave all of us thinking, "How on earth did anyone think of that?"

and "Why does that work?" Stewart James is a wonderful example. He spent a lifetime inventing such magic. A few of his tricks have made their way into basic, widely performed magic. Essentially none of them have been thought about, much less understood. There is gold in these hills, we promise you!

James was born, lived, and died in the tiny town of Courtright, Ontario, in Canada. He was a prolific contributor to magic literature, publishing a dozen books and hundreds of journal articles. Anyone interested in mathematical magic has to come to terms with Stewart James.

We met Stewart at a magic convention in 1961—after hitchhiking from New York to Colon, Michigan, with a friend. Colon is a tiny town and they deal with the hundreds of conventioneers by putting them up in local people's homes. We slept on floors, but, really, we didn't sleep. We talked, watched, performed, and learned magic until we dropped.

Stewart was a respected senior citizen by then and we were kids, but, the way magic works, we were all in the same space. Trying to fool and educate each other, enjoying the rare possibility of talking openly about secrets. A few years later, we wrote to Stewart, trying to make contact with one of the great magicians. The first page of his answer is reproduced in figure 60. It gives more details of magic convention life.

Stewart's letter went on for another page and then stopped abruptly in midsentence. We wrote back, asking if a page was missing and were told that he didn't like to waste time with the usual "Apologies for not answering sooner," "Hello," "Goodbye," etc. Indeed, his next letter began with "pg 3, paragraph *a*," and continued the incomplete sentence. Stewart referred to his letters as "the letter," and continued in this way throughout the correspondence.

After a few years of correspondence, we decided to visit Stewart. This was no simple matter as at that time the closest public transportation stopped about fifty miles from Courtright. Coordinating bus schedules, Stewart picked us up at an appointed time and drove us to his house. We slept in a downstairs bedroom. Stewart cooked, and occasionally left for an hour to do his "rounds," but mostly we talked magic.

The talks were about people, history, and ideas. Stewart had a very good library of magic books that he kept track of with a typewritten, alphabetical index of all his books. Since about one to two hundred books come out each year, this index amazed us—how did he keep it up to date as books accrued? The answer is simple: Every six months or so, he simply sat down and retyped it!

April 23rd, 1968

Page One

From: Stewart James
      1506 St. Clair Parkway
      Courtright, Ontario, Canada

To: Persi Diaconis
      136 W. 11th Street
      New York, N. Y.

(a) A really great pleasure to hear from you.   Probably thought
I wasn't going to answer. Will always reply to a letter from you
but it may not be promptly as in this case.   At present am involved
in completing Volume Two of the Rope ncyclopedia as quickly as
possible.   Sid Lorraine is doing ALL the illustrations this time
and he is tied up on a previous co it ont for a couple of weeks
and so I am taking advantage of this hiatus.

(b) O.K.   This is your address.   But should I not have a Zip Code
Number?

(c) I remember very well meeting you at Colon.   I followed you
around to see Kane's Wild Card and the Egg Bag over and over again.
I remember you doing a card location for the late Stewart Judah.
Probably the greatest compliment I can pay you is to say you were
the most exciting new personality in magic (to e at least) since
Winston Freer.   Entirely different reasons but nevertheless true.

(d) This is not to belittle your performing ability, which is
fantastic, but I was even more impressed with your knowledge.   You
have probably had the experience I have had so any times.   The
individuals who seem to think the beginning of magic was when THEY
became interested in it.   I remember so well you coming down the
aisle to where I was seated during an intermission in one of the
evening shows and talking of Hofzinser.   You gave e the uncanny
sensation of conversing with an old man in a young person's body.

(e) Haxton wrote me one time that Peter Warlock had seen you perform.
PW said you were absolutely amazing.   I told Haxton I was disappointed
PW wasn't impressed more than he was.   By now you must begin to
realize I am a fan of yours.

(f) The article you are looking for would be "STRANGERS FROM 2 WORLDS"
New TOPS, April, 1963, p41.   Like Yates, Hummer's mathematical monte
was the thought starter but in a rather indirect way.   I acquired
the Hummer trick, filed it and thought no more about it.   Then Al
Koran began getting publicity with it and had Stanley market his
presentation.   I took another long look and decided I wanted to
eliminate two things and add a third.   I wanted to work without
a marker.   I wanted to work without EVER seeing the objects.   And
I nted to increase the number of objects.   In SF2W there is no
limit to the number of objects but 5 or 7 at the most seem sufficient.

Figure 60. Letter from Stewart James (1968)

After about a day and a half of talk, we wanted to show Stewart a card trick with his own deck. Up to this time, all the tricks were done "in the air" by talking them through rather than performing them. We asked to borrow a deck. He answered, "This may sound strange but I don't have a real deck of cards in the house—haven't had one for four or five years." Keep in mind that all this time Stewart was producing a monthly column of card tricks and contributing dozens of other articles to various magic journals. He explained: "After all, when Agatha Christie writes a murder mystery, she doesn't have to go out and kill somebody."

Stewart was the constant caretaker for his aging mother. During our visit she was not happy he was sharing his attention and often screamed loudly. He said it was difficult to leave home for more than an hour or two. Stewart's job was as an auxiliary free-delivery mailman. He did this in his own car and had learned to drive on the "wrong side," sitting on the right-hand side of the front seat, steering with his left hand stretched over, his left foot operating the gas and brake pedals. We went out with him one morning, sitting in the back. There was not a lot of traffic on those roads but the occasional passing car seemed shocked to find a driverless vehicle tootling along. One driver followed us for a mile or two. When he got too close, Stewart pulled over so nothing could be seen.

One of Stewart's most famous tricks is called Miraskill. Here is a bland description: Shuffle a full deck of cards and turn over two at a time. If the pair is mixed—one red, one black—they cancel each other out and are set aside. If both are red, the pair is put aside on the left. If both are black, the pair is set aside on the right. At the end, no matter how the pairs are shuffled, the number of red-red pairs is the same as the number of black-black pairs. One may patter as if at a basketball game—"Two points for the red team," etc. The game ends in a tie. If four red cards are removed before the shuffle, the black team will win by four points.

The version Stewart published is almost as bland as just described. There are many variations possible and we wondered why it had not been developed. He explained that he had been thrilled by the principle when he first discovered it and made the mistake of performing it for some jerk who promptly began sending it around, claiming poor, minor variations as his own invention. Stewart sent the bland version to Theo Anneman for publication in *Jinx* magazine to receive some credit for his invention. He told us: "It was stillborn for me after that and I vowed not to think about it."

We have occasionally taught a class on mathematics and magic tricks. One year, in our second meeting with him, a Harvard freshman, Joe Fendel, offered the following version of Miraskill—it is a fine illustration of the gold waiting in Stewart's work. We will quote Joe's own description of the effect:

You may be familiar with the game Rock, Paper, Scissors. If not, don't worry, the rules are simple. Two people randomly choose one of three objects: a rock, a sheet of paper, or a pair of scissors. They simultaneously reveal their choice. If both have chosen the same object, the game is considered to be a tie. If they have chosen different objects then the winner is determined as follows:

If rock and scissors are chosen, the rock is considered to have smashed the scissors, and whoever chose the rock wins.

If scissors and paper arc chosen, the scissors are considered to cut through the paper, and whoever chose the scissors wins.

If paper and rock are chosen, the paper is considered to en-wrap the rock, and whoever chose the paper wins.

So as you see, under random probability, each is equally likely to win and cqually likely to lose. Furthermore, under the usual playground rules, there is no reward for winning, but rather a penalty of some sort to be inflicted on the loser by the winner!

I have provided you with a deck of cards (see figure 61). You will notice that there are exactly twenty-seven cards. Cut the cards and mix them. Holding them face-down, draw one and put it elsewhere, not knowing what it is. Mix the cards again for good measure. Take out a pen to keep score on the score sheet I've provided (see figure 62).

Begin with round one. Go through the deck, drawing two cards at a time. (You will make thirteen draws per round.) For round one, look at the *top* picture on the two cards. If they are the same, it's a tie, so put them in the discard pile. If they are different, score a penalty point for the loser. For example, if the top pictures of the two cards show rock and paper, score one point under rock because paper beats rock. Again, remember:

Rock beats scissors,   paper beats rock,   scissors beat paper.

After all thirteen draws, do a final comparison. Two of the scores should be of the same parity. (Two are odd and one is even, or two are even and one is odd.) Take the two scores that are the same parity, and examine the objects who hold those scores. Place the winning object of the two in the "Final Winner's Box" for round one. For example, if scissors and paper both had odd scores, you would write "scissors" in the "Final Winner's Box," because scissors beat paper.

Figure 61. Joe Fendel's twenty-seven cards

	Rock	Scissors	Paper	Final Winner's Box
Round 1				
Round 2				
Round 3				

Figure 62. Score sheet

Now repeat the process for round two, this time looking at the *middle* picture of each card. After you determine the final winner of round two, repeat the process for round three, this time examining the *bottom* picture.

You should now have your three final winner's boxes full. Concentrate very hard on that column, and fetch the card you initially removed from the deck. Turn it over, and . . .

Lo and behold!!! The pictures on the card *match* the pictures in the final winner's boxes *exactly*.

We leave it to the interested reader to understand just why this works. It is a close cousin to Voodoo Fortune Telling, a trick of Bob Hummer's that also exploits the Miraskill principle.

Martin Gardner had a favorite Stewart James trick—this involved Fibonacci numbers, a sequence where the next number is the sum of the previous two (1, 1, 2, 3, 5, 8, 13, 21, . . .). These numbers were introduced in Fibonacci's book *Liber Abaci* in 1202! The Fibonacci sequence has many variations and applications. We have used it in our "day jobs."[13] One of the most celebrated problems in modern mathematics, Hilbert's tenth problem, was cracked using exotic properties of Fibonacci numbers.[14] There is even a journal, *The Fibonacci Quarterly*, that has collected together the strange and surprising properties of these and related sequences for nearly fifty years.

Additionally, the Fibonacci numbers have a crank aspect to them with ties to the divine proportion and aesthetics. Some of the founders of the *Fibonacci Quarterly* parked only in spots labeled by Fibonacci numbers. There are so many properties just under the surface that they are repeatedly rediscovered by amateur number sleuths. A marvelous account of "Fibonacciana" is in Martin Gardner's *Mathematical Circus*.[15]

Gardner was surprised to find that Stewart had stumbled upon a new property and managed to harness it for a performable trick. We offer a novel variation, an explanation, and a host of generalizations. Here is our version of Stewart's trick.

THE MYSTERIOUS NUMBER SEVEN. The performer shows a blank four-by-four grid and says he has written a prediction on its back (figure 63). If you get a piece of paper and draw such a square, you can follow along and fool yourself as we go. "They teach kids the craziest things in school nowadays. My daughter came home and told me about clock arithmetic. And she was doing it modulo seven! Here, let me show you this. Name any two small numbers." Suppose the spectator says "five" and "three." "I'll start off with your numbers" (the performer writes "five" and "three" in the first two spots in the square). "Now, we'll add 'mod seven.' This just means if the sum is eight or more, we subtract seven. Let's see, five and three are eight. This is more than seven, so we'll subtract seven from eight to get one." The performer writes "one" in the next spot. "Now we continue: Let's see, three and one are four, that's okay, one and four are five, which is okay, four and five make nine, so we subtract seven to get two, five and two are seven, which is still okay, seven and two are nine to get" (the performer may ask for help or ask what the next step is) "two." This

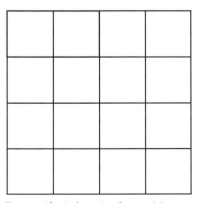

Figure 63. A four-by-four grid

5	3	1	4
5	2	7	2
2	4	6	3
2	5	7	5

Figure 64. Filled-in grid starting with 5 and 3

pattern is continued until all sixteen squares are filled in, resulting in the grid shown in figure 64.

If you are working along with us, start with any two small numbers (but not both sevens!) and fill in the squares, always adding the last two mod seven to get the next number. The surprise comes at the end. With a truly random start and the chaotic pattern filled in, the *ordinary* sum is always exactly *sixty-three*. The performer turns over the grid and shows this has been predicted.

Stewart's original version (the AAG principle)[16] involved a five-by-five grid with addition modulo nine. The performer picks the first number, the spectator picks the second number, and the total is always the performer's choice plus 117 (so if the performer chooses five, the total is 122). The four-by-four Fibonacci mod seven grid (as well as a three-by-three version) was explored at the end of Card Colm's column in June 2007.[17]

How do such tricks work? Let's consider the mod seven version. Here, the spectator can name any two numbers to start (which we will require to be no larger than seven). Thus, there are exactly forty-nine different starting possibilities. However, one of these is ruled out, namely, when the first two choices are both seven. (In this case, filling in the grid by our mod seven rule would result in *all* the squares filled in with sevens, a not very interesting pattern!) Suppose the choice is 1, 2. Then the numbers filled in are 1, 2, 3, 5, 1, 6, 7, 6, 6, 5, 4, 2, 6, 1, 7, 1, 1, 2, 3, 5, . . . , where the dots mean the sequence repeats cyclically with period sixteen, forever. If *any* of the sixteen pairs in this sequence had been chosen, e.g., 2, 3 or 6, 6, then the *same* set of sixteen integers would appear. Hence, in this case the sum is fixed (and, you can check, it is 63). Take a pair that doesn't appear, e.g., 1, 3. This generates 1, 3, 4, 7, 4, 4, 1, 5, 6, 4, 3, 7, 3, 3, 6, 2, 1, 3, 4, 7, ... This gives another set of sixteen pairs and, again, a sum of 63. Finally, a start of 1, 4 gives 1, 4, 5, 2, 7, 2, 2, 4, 6, 3, 2, 5, 7, 5, 5, 3, 1, 4, 5, 2, . . . and, again, a sum of 63. This gives 16 + 16 + 16 = 48 pairs, and the forbidden 7, 7 makes for all forty-nine starting pairs.

In Stewart's version (modulo nine), there are eighty-one possible starts. Calculations show that there are three cycles of length twenty-four (with starting values 1, 1; 2, 2; and 4, 4), one cycle of eight (with starting value 3, 3), and one cycle of length one (with starting value 9, 9). Stewart originally did not allow both of the starting values to be multiples of three (easily achieved by having the spectator *not* choose a multiple of three), in which case the twenty-five numbers in the filled-in grid will be all the numbers in one of the length twenty-four cycles together with a repeat of the first number in position twenty-five (which was just the spectator's chosen number). Since the sum of all twenty-four numbers in any of the "long" cycles is 117, then the predicted sum will be 117 plus the spectator's choice. This version was modified by James who noticed that if the starting pair comes from the "short" (length eight) cycle, then the predicted sum should now be 81 plus the spectator's choice.

There are two key features: The period of the possible starts and the sums of all the numbers in the various orbits. In the modulo seven case, there are three orbits of size sixteen (and one orbit of size one). All the length sixteen orbits have the same sum, 63. One may try various moduli, searching for these two features. For example, modulo five, there is one orbit of length twenty (with a sum of forty) and one orbit of length four (1, 3, 4, 2). Thus, a four-by-five grid can be prepared and a prediction of 40 written. Have the spectator select two numbers and write them down (to avoid the starts 1, 3; 3, 4; 4, 2; and 2, 1 by reversing the order when entering them, i.e., using 3, 1; 4, 3; 2, 4; or 1, 2). The curious reader will find modulo eight gives a mess!

There has been a fair amount of mathematical study of the periods of Fibonacci sequences. For example, it is known that if the modulus is a prime p that is ± 1 (mod five), then the period must divide $p - 1$. Thus, $p = 11$ has a period of 10 (which certainly divides $p - 1$). For a prime p which is ± 2 (mod five), the period must divide $2p - 2$. Much more is known, some of it involving sophisticated results from number theory.[18] To be honest, we do not find these tricks useful magically. They are too slow and too prone to error. Perhaps the reader can create a story that makes them play well.

We end our recollection of Stewart James with his playful side. He loved puns and word games. Two that he showed us may amuse the reader. They start with the well-known arrangement $\overset{wood}{Ralph}$ (read as Ralph Underwood). Can the reader parse $\mathbf{u}\,{}^{all}_{now}\,\mathbf{s}$? What about $B\,\overset{0}{e}\,D$?

CHARLES THORNTON JORDAN

Charles Jordan was the first great inventor of mathematical card tricks. We have devoted several chapters to developments of just one of his ideas: He was the first to use the principles behind de Bruijn sequences

Figure 65. Charles Thornton Jordan

in his trick Colouria (developed here in chapters 2, 3, and 4). Jordan has a mysterious legacy among avant-garde magicians today. He appeared in 1915 "out of nowhere" via advertisements for amazing new tricks in magic magazines. There was a steady flow of new magic, books, and individual secrets until 1923. Then, he essentially disappeared, except for collections of his previous work peddled by entrepreneurs. His magic was sometimes said to be invented by others; he is frequently described as a Petaluma chicken farmer; he invented and sold fancy radios. . . . Here is what little we know.

Jordan was born in Berkeley, California, on October 1, 1888, to Mr. and Mrs. Charles Ronlett Jordan. He died in Petaluma, California, at the home of his sister, Mrs. George Woodson, on March 24, 1944. He began magic as a hobbyist but was too shy to be a reasonable performer. He turned to selling magic to other magicians. One of his most celebrated tricks came into the world this way.

New Ones on the Magical Horizon

Long distance mind reading.   You mail an ordinary pack of cards to anyone, requesting him to shuffle and select a card. He shuffles again and returns only **half** the pack to you, not intimating whether or not it contains his card. By return mail, you name the card he selected.

Price—$2.50

Note: On receipt of 50 cents, I will give you an actual demonstration. Then, if you want the secret, remit the balance of $2.00.

This ad appeared in the premier American magic magazine the *Sphinx* in May 1916. The *Sphinx* sold for ten cents a copy in 1916. Comparable journals sell for five dollars a copy as we write this in early 2009. That makes a *factor* of fifty, so Jordan was selling his trick for a fantastic sum. It is also a fantastic trick; nothing remotely like it had been seen before. Magicians scrambled to find Jordan's magic. One would buy a trick and then type up copies and trade it for others. The magic caused an excitement that carries through to today. We will explain Jordan's trick later in this section but for now we return to his life's story.

Jordan moved to Penngrove, California, and, indeed, worked there as a chicken farmer for many years. This was mixed with many other interests. For a while, he designed and built large radios. There are still some of Jordan's marvelous radios lurking in attics in the Petaluma

Figure 66. Jordan's business card

area. The one we saw was about three feet deep, five feet wide, and three feet high—a massive object, finished in fine woodwork, with huge vacuum tubes inside.

Jordan had an active secret life in another dimension. According to J. Burnham he was "the champion of puzzle contesters of America in the 1920's and 1930's."[19] Newspapers often had contests to increase readership. There was a daily set of trick questions; perhaps a cartoon or anagram hiding a famous name. Such a puzzle is often called a rebus. For example, figure 67 asks the reader to use the pictures to guess the names of four famous explorers. The solutions were given the following day. In this case, the answers to the first is (Admiral) Byrd. The last is DeSoto. We leave the other two to the reader. The contest that Jordan and Blackledge entered involved a variation. The initial offering was shown in figure 68. The right answer, gleaned from reading the text and sounding out the names, is "Theodore Dreiser." This is only one of a list of ninety and all had to be guessed correctly.

Readers who sent in a perfect set of answers won cash prizes. Jordan and a team of coworkers began entering and winning such contests, first in nearby towns and then nationwide. They made tens of thousands of dollars yearly (a huge sum in those days). They were soon banned and had to find shills to win the prizes on their behalf. Jordan turned to the magic community for this. He would find a prominent magician and write along the lines of, "You may remember me as the inventor of card tricks. . . ." Jordan proposed a deal whereby he'd send all the answers and let the magician keep 10 percent of the prize. We have been fortunate enough to obtain a complete record of one such episode. Jordan contacted the well-known magician J. Elder Blackledge in Indianapolis in 1938. Blackledge was a stage performer, some of whose polished routines have become standards among today's magicians. Several of his tricks are described in the multivolume *Tarbell Course in Magic.* He was an inspiration to young magicians in the Indianapolis area. One of them, Harry Riser, has written about Blackledge's legacy at length.[20]

Blackledge accepted Jordan's offer, and here the plot thickens: Jordan's team wasn't the only group of contest professionals. The initial questions were easy and more than two hundred people submitted correct sets of answers to ninety cartoon puzzles. There was a runoff, and another runoff after that. Finally, only a few contestants had

Figure 67. Famous explorers (redrawn from and courtesy of the *Indianapolis News*)

Figure 68. Rebus no. 68 (redrawn from and courtesy of the *Indianapolis News*)

submitted perfect scores on each set of questions. These included Blackledge, who was getting his answers by telegraph and special delivery from Jordan in California. The remaining contestants were required to come into the newspaper office to take a final test in person! Unquestionably, some of the other winners were also fronting for a syndicate, and were as nervous and inept (at puzzle solving) as Blackledge. Blackledge managed to answer five of the questions and finished second (the thousand dollar prize went to Blackledge, who settled up equitably with the Jordan team). The story evolves as a nail-biter and we have reproduced key parts of the correspondence in figures 69–76.

Jordan wasn't only in the rebus business for the money. According to Burnham "He [Jordan] was an enthusiastic collector of rebus and famous-names material, and of books on all contest subjects. On the day of the night he died he asked his physician to bring him a catalogue of surgical instruments for his rebus collection. Charles Jordan was a devoted hobbyist to the end."[21]

TRAILING THE DOVETAIL SHUFFLE TO ITS LAIR. Jordan's first marketed trick, Psychola,[22] and many later effects depend on the observation that an ordinary riffle shuffle does not thoroughly mix a deck of cards. This can be understood by picturing a full deck with the thirteen spades, in order from ace to king, on top. For the shuffle, the deck is cut into two piles as usual, and the two halves are riffled together. The spades will intermingle with the other cards, but the relative order of the spades is (still) ace through king. If you can't picture this, go get a deck and try it out. The crude idea was introduced earlier by the English magician C. O. Wiliams, who published "Reading the Fifty-two Cards after a Genuine Shuffle" in the September 1913 issue of the magazine *Magic*. Jordan acknowledged Williams's trick and went to town with it.

The way we use this principle now is as follows. A deck of cards is mailed to a friend (or given to a group of spectators while the performer turns away). "Give the cards a cut, give them a shuffle, then give them another cut and another shuffle, give them a few more cuts. I'm sure you'll agree that no human can know the name of the top card. Please take this card off, look at it, and remember it. Insert it into the middle of the deck. Give the cards a random cut and another

Penngrove, Calif.,
January 6, 1938.

Mr. J. Elder Blackledge,
4011 North Meridian St.,
Indianapolis, Indiana.

Dear Mr. Blackledge:

In corresponding with Max Holden, of New York City, he suggested that I write you concerning a contest now being conducted by the Indianapolis News. If not interested yourself, I would appreciate it very much if you could put me in touch with some good reliable person who might be.

The contest I refer to is the "Game of Names," and closes within the next few weeks. One of the requirements is that the entrant be a resident of Indiana. As I have no correspondents in your state, I wrote Holden, who has helped me out on matters of this sort on several previous occasions.

In brief, the set-up is this. Working with a friend in Oakland, Calif. on just this type of contest, in each case working through a local name for the entrant, we both ran our incomes into five figures last year, which naturally means that we succeeded in landing a major prize in nearly every case. Nothing in the rules whatever prevents the local entrant from obtaining assistance from whatever source he desires and the whole thing is perfectly legal and above board.

Our proposition is this: We pay all expenses in advance, and when the contest is concluded, we permit the entry name to retain 10% of the win as compensation for acting as our representative. The contests are usually wound up in short order, and in the case of the NEWS, the commission would be $500 in case of a first-prize win, which we have succeeded in accomplishing more than once.

If you would be interested in such a proposition yourself, or if you could refer me to some reliable friend who would be willing to act as our representative, I would be more than grateful for your cooperation.

At any rate, please hold the matter confidential, and let me hear from you as soon as possible, when, if you or some friend of yours would like to take me up, I will immediately send a remittance to cover expenses, and detailed instructions. As the contest is drawing to a close, the utmost haste is essential, and I enclose a stamped addressed envelope for your reply.
Thanking you in advance for any assistance you can render me, and with all good wishes, I remain,

Box 101

Very truly yours,

Charles T. Jordan

Figure 69. First letter from Jordan to Blackledge

Mr. Charles T. Jordan,   January
Box 101,   12
Penngrove, California.   1938

Dear Mr. Jordan:

Upon my return to Indianapolis I find your letter of January sixth.

I see no reason why I can't accept your proposition. My engagements out of the city I think in no way interfere with this. If you will give me the details I'll do all I can to help.

I shall expect to hear from you in the near future.

Sincerely yours,

4011 North Meridan Street,
Indianapolis,   Indiana.

Figure 70. Blackledge's response to Jordan

Mr. J. Elder Blackledge,   Penngrove, Calif.,
Indianapolis, Indiana.   January 15, 1938

Dear Mr. Blackledge:

    Thank you for yours of the 12th, which arrived this morning. I can't tell
you how much I appreciate your willingness to cooperate with us in this matter.

    First off, I enclose a money order for $10.00, which will more than cover
preliminary expenses.

    The first thing necessary to do is to obtain from the paper back cop-
ies of all pictures, and the answer forms for each week. I believe today is the
close of the first 13 weeks of the fifteen-week contest, and as they charge three
cents apiece for the back pictures, it will be necessary for you to call at the
paper office and purchase all back pictures, or to cut out the form which appears
each day in the paper and mail it to them with the necessary amount of money. You
have them mail them to you, of course, and immediately [after] you receive them,
please mail the entire lot by air-mail special delivery, to Mr. J. S. Railsback,
2459 Truman Avenue, Oakland, California. This is my partner's address, and we use
it for most of the correspondence, as mail connections are so much better there.

    Then it will be best to subscribe to the paper for a month or two, and
clip the pictures each day, and mail same say every two or three days to the same
address in Oakland. It is not necessary to send us the answer forms. Just retain
them, and as soon as we have the puzzles solved we will return them to you, and
you can send in the entire lot of ninety, with the necessary $1.50 during the
week which is allowed following publication of the final picture.

    I am sure everything will work out all right if you will have some member
of your family tend to the mailings to us in case you are out of town.

    Of course these preliminary pictures are comparatively simple. The real
difficulties arise upon issuance of the tie-breakers. Usually anywhere from 10
to two or three hundred contestants submit correct solutions to the preliminary
pictures. Then, in two or three weeks, the paper mails out tie-breakers, which
contain anywhere from 60 to 90 extremely difficult puzzles. It is extremely urgent
that these tie-breaking pictures reach us at the earliest possible moment after
they are issued. We [will] have them photostated and [will] return the originals
to you immediately, sending you the solutions by air mail daily, and the remaining few by wire. As they sometimes allow only five or six days for working the
tie-breakers, you can see the necessity of their reaching Oakland on the earliest
possible plane after they are received by you. So if you are not in town at the
time, if you can arrange to have them forwarded to us by some trusted friend or
relative everything should work out nicely.

    When we send you the solutions we also include a complete explanation
[of] how they are arrived at, so that you will be at all times kept informed as
to what it is all about. Naturally our names etc. are not to be mentioned to the
paper, but after the whole thing is over there is nothing to be harmed by admitting that you had outside assistance if the question is ever brought up. It seldom
is, as nothing in the rules declares that you may not have all the assistance
you desire.

    As I wrote you previously, we will pay all expenses in advance, and when
and if a prize is won, you are to retain ten percent of it and your state income
tax, if your state has one, [with] the balance to come to us -- we of course pay
the federal income tax of same. But all of that can be gone into later, of course.

    Mr. Railsback will write you further on this matter just as soon as we
know that you have mailed the pictures to us. And I would appreciate it very much
if you will wire him collect, in Oakland, just as soon as you have mailed the
puzzles, to some such purport as:

    "Papers mailed to-day."

    Thanking you again for your assistance, and trusting that the whole en-
terprise will prove mutually profitable to us, I remain,

            Sincerely,

Box 101   Charles T. Jordan.

Figure 71. Jordan's response to
Blackledge

Dear Contestant:

This is Official Notification from The Indianapolis News that as a result of your answers submitted in the "Game of Names" Contest, you are tied for a prize.

Just what prize you will win, if any, will depend on your answers to the tie-breaking cartoons we are enclosing. This is in accordance with Rule No. 5 of the contest which provides, in case of ties, another set of cartoons will be submitted to the persons tied.

Your answers to this tie-breaking series of cartoons must be filled in directly under the cartoons to which they apply, and the folder must be returned or mailed to The Indianapolis News on or before Midnight of ___MAR   6 1938___ .

Trusting that we will receive your answers promptly, and with best wishes for success, we are,

                              Cordially yours,
Enc.

                              Games of Names Editor
                              THE INDIANAPOLIS NEWS

Figure 72. Newspaper informs them that they are eligible for the tiebreaker

LD65 HG 14
          INDIANAPOLIS IND 1234P MAR 25 1938
     J ELDER BLACKLEDGE, DELIVER PERSONAL ONLY
          4011 NORTH MERIDIAN ST INDPLS
     OFFICIAL NOTIFICATION THAT YOU MUST WORK TIEBREAKER
     AT OUR OFFICE NINE AM MARCH 26
          CONTEST DEPARTMENT THE INDIANAPOLIS NEWS
                                        1252P

Figure 73. Newspaper informs Blackledge that he must appear in person for the tiebreaker

2459 Truman Ave.
Oakland, Calif.,
March 10, 1938.

Dear Mr. Blackledge:

      In spite of the fact the mail planes were delayed last week end, we believe we had all titles correct on the breaker. Another breaker came from Boston at the same time but we were unable to complete it, so had to drop out.

      Inasmuch as there may possibly be another breaker to solve, and you couldn't mail us your own copy, due to the short time they would probably allow, our only chance to stay in the running, if another breaker is necessary, is to have you get a photostatic copy of it and shoot it out. So I am enclosing $15.00, which should be enough to cover the cost, and trust you will do so if necessary. Any Blue Print shop will make it for you, and will usually hurry it if you ask them to. We would want one negative and one positive copy, the same size as the original. If we can get even twelve hours on it, we will have a good chance to crack it. If this becomes necessary, please wire me the same as before, giving number of pictures, closing date, and the time you mail it, so we can figure when to expect it.

      You might phone the contest editor and ask him when the results of the first tie-breaker will be announced, and if any further tie-breakers will have to be solved. Might get an idea that way as to the possibility of another one.

      If they should call on you and tell you that there were ties for first place, and ask if you would rather divide up the prize money involved than work on another one, tell them by all means you would be willing to split. This would insure a fair prize and avoid any further work on it. This often happens when only a few are tied

      You have probably grasped the way in which these puzzles are solved, that is, the title to a given cartoon is made up of the sounds, syllables, or words represented by the queer objects, etc. in the picture. For instance, picture No. 4, which was "Tom and Jerry". TOMAN (Scotch) is a mound (the word mound, uttered by a Scotchman, appears in the picture)--D, the letter is on a sign post--JERRY (a railroad section worker). Running these together, we get TOMAN-D-JERRY, or Tom and Jerry. It's just a matter of dividing up each possible title under the picture, until we find the definitions that fit words under the picture, which strung together, produce the complete title. Just plain old dictionary digging, and you'd be surprised the number of different ways a word can be spelled and still pronounced the same way.

      If you get any advance information from the paper, please let me know at once. And many thanks for your efficient and kind help so far.

                   Sincerely,
                   J. S. Railsback

Figure 74. Jordan's partner, J. S. Railsback, informs Blackledge they are in the tiebreaker

Mr. J. S. Railsback,   April
2459 Truman Avenue,   6
Oakland, California.   1938

Dear Mr. Railsback:

I have been waiting to write you until I know what the results of the contest were. In the yesterday afternoon News the enclosed was published.

As you know, on March 26th I went down to the News Office with four others
for the second tie-breaker. There were two women (Mrs. O'Hara was one) and
two other men. We worked from 9:00 in the morning until 6:00 that afternoon. From the check on the answer list I got five correct - which was not
quite enough. These were, to me at least, very difficult. I did the best I
could going in there cold - just about the same as playing tennis with Bill
Tilden. It was quite a surprise to me to be called but I didn't want to let
you down. I had never solved any of these before and all I knew about it was
the instruction Mr. Leonardgave me. And by the way, none of the tips were
there. Of course, had you been able to solve this last bunch yourselves the
result would have been first. I am sorry I fell down on you but, as I say,
I did the best I could under the circumstances.

I have the check and as soon as I find out just what my taxes on the full
amount will be I'll send the balance to you. Do you want the check made
out to you?

With Best regards, I am
Sincerely yours,

Figure 75. Blackledge describes the in-person tiebreaker to Railsback

Penngrove, Calif.,
May 10, 1938.

Dear Mr. Blackledge,

      It may seem like criminal negligence, my not having written you
sooner, but for the last few weeks we have been tied up with a series of tie-
breakers of various sorts that has simply kept us on the jump every minute.
      Wanted to thank you sincerely for the fine attention you gave the
Indianapolis contest, and for your swell sportsmanship in daring the brave
the interior of the newspaper office for that totally unexpected tie-breaker.
Usually when so few are tied the paper goes around to see them, and ask if
they care to split the prizes, so naturally we anticipated that procedure if
another tie-breaker was not mailed out.
      Anyway, it's over with now, and I thank you again, also for your
prompt remittance. If I had any favorite card trick or a dozen of them, they
would be yours instanter [sic], but the truth of the matter is, I gave up all
connection with magic years ago, and have practically nothing pertaining to
it on the place.
      Railsback and I will be very glad to see you when you come to the
coast this fall. Be sure and advise us well in advance.

                              Sincerely,
Box 101
                              Charles T. Jordan

Figure 76. Jordan congratulates Blackledge on his in-person performance

shuffle. Mail the deck back to me. Oh yes, at 6:00 p.m., every evening, please concentrate on your card." When the performer gets the deck back, it is possible to decode the mess and find the selected card.

The deck the performer mails is arranged in a known order. For the purpose of this description, let us suppose that the spades are in order on top, followed by the ace through king of clubs, then the ace through king of hearts, and finally the ace through king of diamonds. As explained above, a single riffle shuffle leaves the cards in two interlocking chains (figure 77).

What happens after a second shuffle? The top half is cut off—this has two chains. The bottom half also has two chains. When the two halves are riffled together, the deck has four chains. A further shuffle results in eight chains. When the top card is noted and moved to the middle, it makes a ninth chain of size one!

When the performer gets the deck back, the chains can be undone by playing a kind of solitaire. Turn up the top card and deal it face-up on the table. Say it is the six of hearts. If the next card is the seven of hearts, play it on the six. If not, start a new pile. Continue this, starting new piles, with subsequent cards being played on previous cards. As you go through the deck you will form eight piles, each of size about one-eighth of the pack, and a ninth pile, of size one containing the selected card.

Since the cards have been repeatedly cut they must be treated cyclically, so the bottom card (the king of diamonds) is followed by the original top card (the ace of spades). It is best to practic this with cuts followed by one shuffle, then two shuffles. The trick has a small chance of failing (if the selection is placed back very close to the top, for example). Extensive empirical experiments, as reported by D. Bayer and

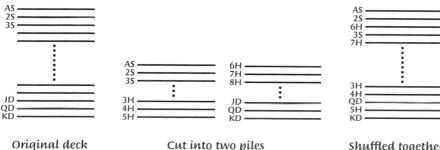

Original deck   Cut into two piles   Shuffled together   Figure 77. Two chains

P. Diaconis,[23] show that the trick works more than 90 percent of the time. The reference also contains a more refined way of undoing the eight chains.

Jordan's trick was the key to one of our most celebrated mathematical findings—the theorem showing that it takes at least seven ordinary riffle shuffles to adequately mix up fifty-two cards. The discussion above shows that three shuffles are definitely not enough. Similarly, after five shuffles, the deck can have at most thirty-two chains. On average, a well-mixed deck has about twenty-six chains, but can easily have more than thirty-two. A more careful analysis shows that, after five shuffles, a sizable proportion of the possible arrangements still can't be reached. Note that this is true no matter how the shuffles are performed: Neatly, randomly, or dependent from shuffle to shuffle. To go further, we used a natural model of riffle shuffles introduced by Bell Labs scientists Ed Gilbert, Claude Shannon, and Jim Reeds. This shows that the original order fades away starting at seven shuffles, tending to zero exponentially fast. The details are too technical for the present treatment.[24] The analysis of shuffles has many further mathematical implications and many practical implications as well.[25]

Returning to magic, Jordan's original Long Distance Mind Reading trick took no chances. The deck was repeatedly cut, riffle shuffled once, and the cards were then cut into two piles. A card is randomly chosen from the middle of one of the two piles, noted, and moved to the opposite pile. Finally, either pile is chosen, shuffled thoroughly, and mailed to the performer. With the information above, the reader should be able to see how the performer can take the shuffled pile and determine the selected card. It takes a bit of thought but is relatively surefire.

We have accrued a lot of mileage from Jordan's inventions, both magically and mathematically. He has inspired all who are serious about mathematical magic. Stewart James wrote, "To me, Jordan was the greatest idea man that card magic has produced."[26] As we write this, we have Henry Christ's file of Jordan's individually sold secrets open in front of us. Bought one at a time as they came out, they were lovingly read and saved (both the envelopes and the faded "Secret Gimmicks"). The individual secrets sold from fifty cents to two dollars or so in the 1920s. There were about a hundred of them. Again, this makes for a sizable investment. Looking back, we are glad it was made and wish there were more of these saved secrets to be found.

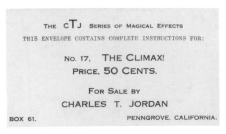

THE cTj SERIES OF MAGICAL EFFECTS
THIS ENVELOPE CONTAINS COMPLETE INSTRUCTIONS FOR:

No. 17,   THE CLIMAX!
PRICE, 50 CENTS.

FOR SALE BY
CHARLES T. JORDAN
BOX 61.   PENNGROVE, CALIFORNIA.

Figure 78. One of Jordan's envelopes

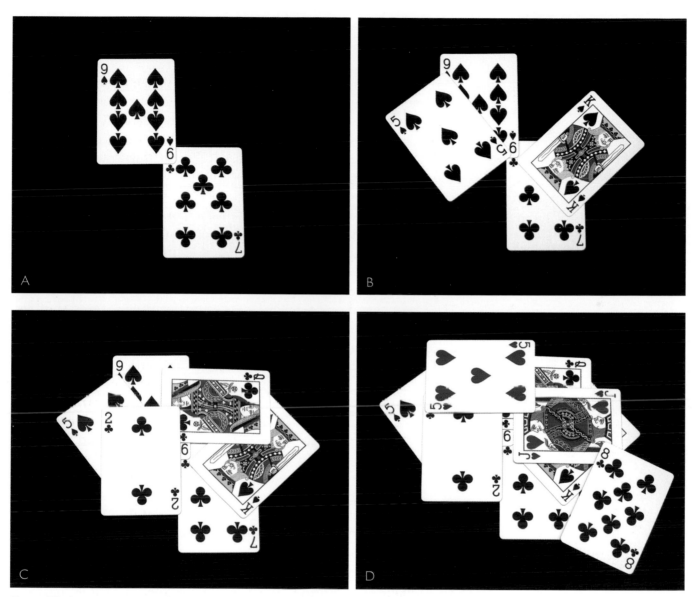

Figure 79. An arrangement of cards (see figure 80 for an explanation)

BOB HUMMER

The word genius gets tossed around easily. We have not used it else-where but, for us, Bob Hummer was a genius at inventing mathematical magic. He was also as odd and original a personality as we have encountered. We began this book with his "turn over two" ideas. His

> THE CLIMAX!

PRELIMINARY PREPARATION:— Remove the four sixes from a blue-backed pack of cards and place them in four of your pockets, remembering which is in each pocket. Find the four Sevens and the four Nines. Place the Sevens at four spots on the table, each face up and seven or eight inches distant from the others. We are going to give the Sevens such an innocent appearance that any one would swear they were Sixes — indexes and all. You will be instructed how to make one Seven into a Six, the same operation holding good for the other three. The accompanying illustrations make the matter clearer than a page of words could. Note that you first place the corner of a Nine of the right color over an index of a Seven, THE REVERSED "9" ("6") making a perfect "SIX." (See 1.) Then place additional cards, as in 2, which conceal the juncture of the two. Additional cards are applied, as in 3, and a few more as in 4. Could you desire a more natural appearing Six-spot than the camouflaged Seven in 4? Each of the Sevens is built up in this manner, and balance of the pack is scattered face up over the table, connecting the four foundational heaps and giving the appearance of a pack scattered haphazard, face up on the table. NOW remove the four Sixes from a red-backed deck, and put them in a heap face up, just in front of the layout on table. They seem to be a part of the scattered blue pack, if they are observed at all.

PRESENTATION:— Pass the red-backed pack (minus the Sixes) for a thorough shuffle, and holding it in plain sight, place it face up on table, directly on the four Sixes. Step away. Ask a spectator to step to table. He turns the red pack (now complete) face down in the clear. He then deals pack, one card at a time, into four face down heaps. This brings a Six to bottom of each heap. Turn away, and ask him to lift any heap and note card at face of it — then to gather up the four heaps, shuffle pack and lay same aside. Now direct him to look at the scattered pack, and tell you if a duplicate of his card is there. Naturally he replies: "Yes," for it seems to be. Have him place his two hands flat on the cards, that none may escape. Spread a borrowed handkerchief over his hands and the cards, and have him square up pack under its cover. This done, he removes the handkerchief himself. Tell him to grasp the cards very tightly, and to name card he chose when you count "Three!" He does and to his amazement, you (reaching into the proper pocket(produce his card therefrom. Then watch his face as he searches the pack.

Figure 80. Instructions for *The Climax*

mathematical Three-Card Monte is developed below. Let us relate what little we know of his life.

Hummer was born on January 25, 1906, and died in Havre de Grace, Maryland, in April 1981. His father worked for the Salvation Army. He had a brother and sister but we do not know many details of his early life. He first appeared on the magic scene as a secret assistant for a great stage card worker, Paul Le Paul. Le Paul would call Hummer up from the audience as a "volunteer." Hummer had a very funny persona on stage that just naturally made people laugh. While we will focus on his mathematical inventions, Hummer also did fine, original sleight of hand (we still use his one-handed color change). He invented and performed strange tricks. He could toss a single card up into the air so it made a 360-degree circle around his body before being caught. One spectacular feat—after a card was selected and signed, he threw the deck at a nearby window with a half-open window shade. The shade sprang up and the spectator's card was seen stuck to the window. When the spectator went to check the signature, shock set in. The card was stuck to the glass *outside* the window.

Hummer made a living doing magic around Chicago, often busking in bars and passing the hat. He hung around magicians, swapping tricks and telling awful jokes: "A family of skunks sat down for dinner around a head of lettuce. Bowing heads, the father said 'Let us pray'" (= lettuce pray = let us spray). He performed throughout the Midwest, never a star and never driven to make money. He worked when he needed to.

Figure 81. Two publicity photos of Bob Hummer

Figure 82. A notice for Bob Hummer's act in 1942 (from the *Billboard*)

Along the way, he began to market easy-to-do but highly original magic. Perhaps his most well-known trick is called Mathematical Three-Card Monte. First published in 1951, and widely varied and performed since, the trick fools you even after you know how it is done. It uses the fingers as a true digital computer and has a Lewis Carroll logician's flavor. We begin with the original, then pass to standard variations. Then, we show how a look at the mathematics shows that an improvement is available. Finally, we provide some unpublished variations by the movie director (and great card man) Cy Enfield.

HUMMER'S THREE-CARD MONTE

THE ORIGINAL. Three random cards are placed face-up on the table. The performer says to the spectator they are going to play "mental three-card monte" and turns away. He asks the spectator to switch various pairs of cards to mix things up, then think of one of the three cards and make a few more switches. Without turning back around, the performer says: "Concentrate on your card . . . let's see; please pick up the card on your left"—this proves to be the chosen card.

WAIT, WAIT. The above is the *effect* as the spectator sees it. There are, as always, pertinent details to be filled in. First, as the spectator makes the initial switches, they are called out, as in "left and middle" or "left and right." After a few switches, the performer asks the spectator to think of a card, remember it, and silently switch the other two.

A few more called-out switches are made. Then the performer successfully reveals the chosen card.

As to the method: When the cards are laid out, spot the middle card (say it is the ace of spades for this description). The right hand will act as a tiny computer. To begin, touch the thumb to the middle finger (where the ace of spades is). Turn your back to the spectator and ask him to switch a pair. If he switches the ace, your thumb moves to the appropriate finger. When we do it, the right hand is palm-down and the first finger corresponds to the left card. If the spectator switches "left and middle," your thumb moves to the first finger. If the next switch is "middle and right," your thumb stays on the first finger. If the next is "left and right," your thumb moves to the third finger—always showing where the ace goes. This may be continued for as long as needed, but two or three switches is usually enough.

Have the spectator think of one of the three cards and silently switch the other two. Your thumb stays put. Have him call out some more switches and your thumb keeps track as before, following the presumed ace. At the end, ask the spectator to hand you the card at the position indicated by your thumb (e.g., the left card if your thumb is on the first finger). Glance at it; if it *is* the ace of spades, that's the chosen card and the trick is over. If it is *not* the ace of spades then, without missing a beat, say, "Please hand me one of the remaining cards." Glance at it; if it is the ace of spades, say: "Congratulations, the card on the table is the card you are thinking of." If it is not the ace of spades, then the card just handed to you is the chosen card. Say "Congratulations—I guessed wrong but you are right on the money. The chosen card is [name the card just handed to you]."

As we said, it takes a bit of thought to see why it works, even after you know *how* it works. We leave that pleasure to you.

VARIATIONS. There is no reason to use cards. Indeed, a pricey stage presentation with three brightly colored billiard balls (say, red, green, and yellow) was sold in the late 1950s as Chop-Chop's Mental Colorama. Just listen to this:

For the magician that likes distinction!! The aristocrat of magic props! It's the 1—2—3 of magic mentalism. No. 1—For your desk or den. A conversation piece to be exhibited! No. 2—Do a mental effect simply, clearly and colorfully with masterful results!

No. 3—Just have a spectator think of one of the three perfect lucite colored balls (set on a highly polished small black lucite stand). And while your back is turned, have them move the red, green and yellow balls from space to space any number of times. When they decide to stop, you turn around and immediately announce the color of the selected ball. Mental Colorama has enabled you to be a master mentalist as fast as you can say 1—2—3. The magician in your family will treasure this beauty. **Price—$10.00.**[27]

For the record, Hummer's two-page description sold for one dollar. The cost of the magic journal in which Mental Colorama was advertised was sixty cents. In 2011, *Genii* sells for six dollars an issue. Overblown though it may be, it does show that Hummer's little trick can be built up to something that could be performed for an audience of a thousand.

A popular way to get away from cards was marketed by our friend Al Koran, a British mentalist, in the 1950s. Here, three coffee cups are borrowed and turned mouth-down in a row on the table. The performer (you) notices an identifying mark on one of the cups and follows this cup as if it were the ace of spades. A spectator puts a crumpled-up dollar bill under any one of the cups. Turning away from the proceedings, say you are going to play a mental version of the three-shell game and ask her to silently switch the positions of the other two cups as a practice move. Have her call out a sequence of switches, as in "left—right" or "left—center." To conclude, you turn to face the cups. If the marked cup is in the position indicated by your thumb, that cup contains the bill. If not, the bill is under the cup not in the designated position. It's a good trick and bears repeating a time or two. Koran managed a spectacular finish by introducing some sleight of hand so that he could name the serial number on the bill.

These variations use two pieces of information. For example, the magician must turn around in the bill-under-the-cup version. The New York memory expert Harry Lorayne introduced a neat variation: Three objects, say a coin, a matchbox, and a straw, are placed on the table. A spectator thinks of one, switches the other two, and calls out the switches as before. At the end, the performer asks if the original objects happen to be in the original order. If not, the spectator is asked to call out switches until the order is restored. To conclude, consult your thumb! If it indicates that the original central object is back in the center, that's the object. If not, the thought-of object is at the position not represented by your thumb nor in the center.

Martin Gardner, improving on a clever variation of Hummer's trick by Max Abrams,[28] suggested the following highly performable variation that eliminates the spectator's calling out of switches. It is explained here for the first time, with his permission. Take three cards from a borrowed, shuffled deck and place them face-down on the table. Spot an imperfection (a spot or bend) in one of the three; this is surprisingly easy, particularly with a well-worn deck. Turn your back and have the spectator look at one of the three cards. He then silently switches the other two. You turn back around. If the spotted card is still in its original position, that's the thought-of card. If not, the thought-of card is what's left after eliminating the spotted card and the card currently where the spotted card was. You thus know where the thought-of card is. Now engage in a quick set of switches, like a street three-card monte operator. Keep track of the selected card. To conclude, have another spectator put a hand on one of the three cards. If it is the selection, conclude there (with an appropriate buildup). If not, say "We'll eliminate that card." Have the original spectator choose one (perhaps after more mixing). If correct, conclude there. If not, say "With two cards eliminated, the last must be correct. Name your card." And turn it over to show it is correct.

OUR CONTRIBUTION. In thinking things through, we realized that previous versions do not use all the information. To explain things clearly, suppose an ace, a two, and a three are face-up on the table, ordered left to right. As the spectator calls out switches, there is enough information to follow the exact position of all three cards. Now the spectator makes a silent switch as above (after thinking of one of the three cards). There are only three possibilities as unknowns. After further switches are called out, there are still only three unknowns. To conclude, the performer asks for (say) the middle card. This can be any of the three and gives enough information to determine the thought-of card. If it's what it should be in following the order, it *is* the thought-of card. If not, then neither it nor the projected center card work so the third possibility is the thought-of card.

This eliminates the second question in the original Hummer trick and the awkward "Let's reset the cards to their original order." Finally, it is easy to mechanically follow all three cards. One method is to use both thumbs. The right hand's thumb follows the original center card and the left thumb follows the original left-hand card. If you don't want to use your fingers, use a dodge of Hummers—locate three objects in your view, say a lamp, a clock, and a window. Shift your gaze between the three as the switches are called for the central object and point to one of the three with a finger for the second object. It takes practice, but it's not exactly what anyone would call sleight of hand. Wait, maybe the finger version does count as sleight of hand!

We worked out a variation that might help in some situations. This involves following only one card (say card number one) as with the original. However, you also keep track of parity (even or odd) of the total number of switches. To conclude, ask for the card at the position indicated by your thumb as in the original Hummer presentation. If this is what it should be, stop. If not, the thought-of card is still on the table. It is either to the right (+1) of the removed card, or to the left (−1) of the removed card, going cyclically. Here is a simple rule. If the card handed to you is the two, count +1; if it is the three, count −1. If the number of switches is even, you are done. If the number of switches is odd, switch the signs. This gives the position of the thought-of card.

CY ENDFIELD'S ROCK, PAPER, SCISSORS

Cy Endfield was an amazing guy. A famous American movie director (*Sands of the Kalahari*, *Zulu*, *Gentleman Joe Palooka*) blacklisted during the McCarthy era, he moved to England and made more movies as well as a new life. Cy was a very skillful card handler. The three volumes of his *Entertaining Card Magic* set a new standard for advanced card magic that also provided solid performance material. Later, he created a sensation with a hand-crafted chess set that stacked into two pens but broke down into real playable chess pieces.

He and his wife, Maureen, made the ragtime music of Scott Joplin into a musical (about his hometown of Scranton, Pennsylvania). Of Cy's many successes, we mention his Microwriter—a one-handed keyboard that became a successful product in the early days of the computer revolution.

We carried on an extensive correspondence with Cy over a forty-year period. For about five years we worked on Hummer's Mathematical Three-Card Monte. The following routine, worked out jointly, was in Cy's performing repertoire. He was proud enough to perform a version by mail for Martin Gardner—Martin gave us all of his magical correspondence. We were proud to see that our trick fooled Martin. He asked Cy for the details but Cy turned him down! Here is the secret.

ROCK, PAPER, SCISSORS. The performer explains the classical game of Rock, Paper, Scissors. Each dominates another: Paper covers rock, rock breaks scissors, and scissors cut paper, in a nontransitive cycle. The performer turns his back to the audience and a spectator chooses one of the three objects, then silently switches the other two. A second spectator chooses one of the other two objects and switches them. Finally, a third spectator is stuck with the third object. After a few more switches, without turning around, the performer successfully reveals all three choices.

Here are the performance details. Take a single sheet of paper, tear it into three roughly equal pieces, with one piece slightly larger so the three pieces can be distinguished as large, center (with two rough edges), and small. Write "rock," "paper," and "scissors" on the pieces in a known order (e.g., "rock" on small, "paper" on center, "scissors" on large). Turn your back and have Spectator One think of one of the three and silently switch the other two. Have Spectator Two choose one of the remaining two objects, and switch the other two, calling out his switch (e.g., "left—right"). Ask him to make a few more random switches, calling them out. Spectator Three is asked to concentrate on the third object.

With your back turned at all times, you reason as follows. When the second spectator calls out his switch (e.g., "left—right"), you remember the position *not* called (here, "middle"). This will be the object thought of by the first spectator. They may make any further number of switches. You follow the position of the first player's object.

At the end, tell Spectator Three to pick up the object that was at the position unnamed by the second spectator. Tell the second spectator to pick up the object that is at the position you're following. The first spectator's object is the one that remains on the table. Done with bravado, it seems that you have divined all three choices.

It takes practice to do this smoothly, and some thought to see why it works. Thus far, the fact that the performer can recognize the pieces by shape has not been used. The trick above can be done with any three objects, e.g., an actual rock, a piece of paper, and a pair of scissors (or a coin, watch, and banana). The reason for the shape comes in the next phase.

BOB NEALE'S ROCK, PAPER, SCISSORS. Turn the three pieces of paper face-down, mix them a bit, and place them, one at a time, in front of the three spectators. Because of the shapes, you know who has what. Suppose they are as:

Tom   Dick   Harry
Rock   Paper   Scissors

Again, the performer turns his back to the audience and has the spectators switch places a few times. They don't have to tell you who they are, they just have to call out "switch." Similarly, they can switch pairs of playing pieces, rock and paper, etc. They keep up either kind of switching for a few moves. The effect appears in three phases:

1. After a few switches, the performer (correctly) announces that Tom beats Harry.
2. More switches; the performer asks for the names of two players and then divines who beats whom.
3. More switching; spectators pick one player (e.g., Tom). The performer announces that Tom should play Dick to win.

We leave the working of this to you as a problem in logic. It's actually amazing. This trick is Cy's version of Bob Neale's first trick from *This Is Not a Book.* It is a development of Neale's End Game from the *Pallbearers Review.*[29] Cy dressed the trick up as a battle between three gladiators with a dagger, pike, and net (pike breaks dagger, dagger penetrates net, and net covers pike). He had versions with chess pieces and playing cards. In retrospect, the torn paper dodge is not really needed. Working directly with a small rock, piece of paper and pair of scissors is probably better. For the Neale trick, one can work without props. Have three spectators hold out their hands, one as a fist (rock), one open handed (paper), and one with an extended pair of fingers (scissors). When two switch, they just switch their positions.

This has been an extended development of Hummer's original idea. Our first chapter did the same. It seems to us that many of

Hummer's other ideas have similar promise. The reader will have to work to find them.

MARTIN GARDNER

Just before writing this section we typed "Martin Gardner" into the on-line bookstore Amazon.com. One hundred and forty-five titles came up. It is a remarkable feat to have written (or edited) this many books. It's even more amazing to have a large fraction of that number in print. They include novels, philosophy books, popular science books, books of poems, and riddle and puzzle books. Martin is the dean of popular science writers. A ferocious debunker of the occult and pseu-doscience, a skillful sleight-of-hand worker, but, mostly, he was our friend. For more than fifty years he encouraged and taught us, pub-lished our tricks and mathematics, and was a focal point for the latest and funniest of these.

For twenty-five years, from 1956 to 1981, Martin wrote a mathemati-cal games column for *Scientific American*. This opened the gentle art of recreational mathematics to an international audience of millions. In the course of his column, Martin truly changed the world—he pub-lished the first description of public-key cryptography which is now used in many online banking transactions.[30] He published the first de-scriptions of John Horton Conway's Game of Life and Roger Penrose's celebrated Penrose tiles.[31] The Game of Life so overloaded computers around the world that the game was forcibly banned in many locations.

Martin's column did much more. A blurb that appears on one of his books says:

> **Warning**: Martin Gardner has turned dozens of innocent young-sters into math professors and thousands of math professors into innocent youngsters.

We are living proof; Martin nurtured a runaway fourteen-year-old, pub-lished some of our mathematical findings to give a first publication (in *Scientific American*), found time to occasionally help with homework, and, when the time came to apply for graduate school, Martin was one of our letter writers. There are heart-warming stories here. Mar-tin's letter of recommendation said something like: "I don't know a lot about mathematics but this kid invented two of the best card tricks of the past ten years. You ought to give him a chance." Fred Mosteller, a Harvard statistics professor and keen amateur magician, was on the admissions committee and let the kid into Harvard. Fred became the kid's thesis advisor and, after graduation, the kid eventually returned to Harvard as a professor.

One other tale about Martin's letter. It was sent to a long list of graduate schools. He got a reply from Martin Kruskal at Princeton (a major mathematician who was most well-known for his discovery of solitons) that went roughly: "It's true, Martin. You don't know about mathematics. No one with this kid's limited background could ever make it through a serious math department." Kruskal went on to explain what has come to be known as the Kruskal principle. This is a broadly useful new principle in card magic. A few years later, the kid lectured at the Institute for Defense Analyses, a kind of cryptography think tank in Princeton. Kruskal came up afterwards, full of enthusiasm for the lecture, and asked: "How come I never heard of you? That was wonderful!" The kid tried to remind Kruskal of their history. Kruskal denied it but the kid still has the letter. This was one of the few times that Martin Kruskal's keen insight led him astray!

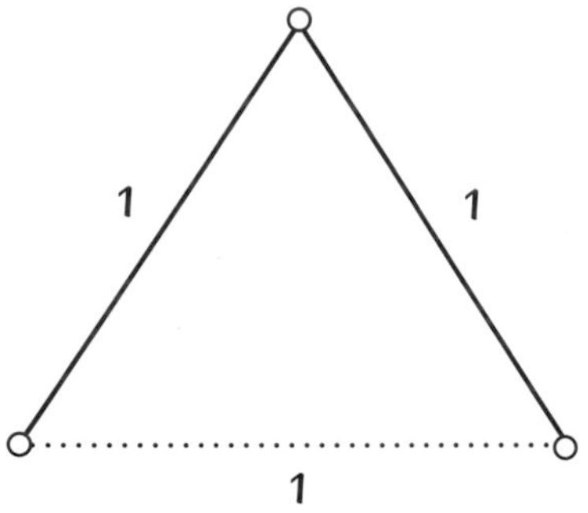

One secret of Martin's writing is that he touches the material he writes about, building small models, trying examples, internalizing, playing with the examples and theorems until he "sees" them. Here is an illustration, and the story of Martin's first research paper.

The story starts in the think tank run by the largest telephone company—Bell Laboratories. Understandably, they study the problem of how to connect a bunch of sites using the least amount of wire (or optical fiber, nowadays). To get a feel for this, consider the three vertices of an equilateral triangle, each side of length one. Running the wire along two of the sides connects the three points together, and uses a total wire length of two (see figure 83).

However, consider adding a fictional point in the middle (called a Steiner point), and connecting all vertices to this middle point (see figure 84). This network still connects all three of the triangle vertices together and its total length is only $\sqrt{3}$. An easy calculation shows that we have managed to save length by a factor of $\frac{\sqrt{3}}{2} = 0.866. \ldots$. This raises the problem of just how much can be saved and how to do it. Is this central point the best? (Yes.) What about more general sets of vertices? Even for the case when we have the four vertices of a square, it isn't obvious what the best configuration is. It turns out that it is the

Figure 83. Three vertices of a unit triangle

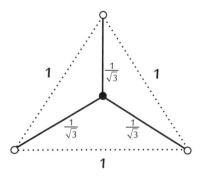

Figure 84. Adding an extra point in the middle

one shown in figure 85 (which can also be rotated 90 degrees). In this case, the total length of the shortest possible network connecting the four vertices of a unit square is $1 + \sqrt{3} = 2.73205\ldots$, no matter how many extra points you try to use! Compare this with the length of three if no additional points are allowed.

Working with Fan Chung, we studied the problem of finding the best network for a set of points on a $2 \times n$ square grid. In particular, we were able to show that when n was an even number, then the best thing to do is to join these 2×2 squares together by single edges (see figure 86), so, in particular, we only had to add n additional Steiner points. However, the shortest network when n is odd needs $2n - 2$ additional Steiner points and has a much more complex structure.

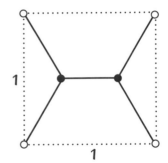

Figure 85. Shortest network for unit square

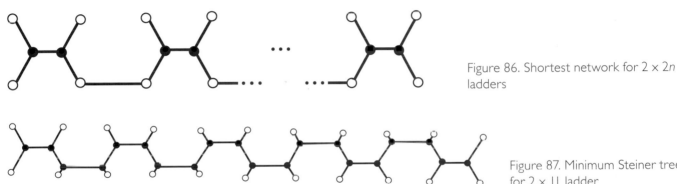

Figure 86. Shortest network for $2 \times 2n$ ladders

Figure 87. Minimum Steiner tree for 2×11 ladder

In fact, it is now known that you can never save more than a factor of $\frac{\sqrt{3}}{2}$ for *any* set of vertices when you are allowed to arbitrarily add additional Steiner points. A beautiful open problem is to find the corresponding best ratio for sets of vertices that are in three dimensions. The conjectured ratio in this case is the amazing quantity:[32]

$$\sqrt{\frac{283 - 3\sqrt{21}}{700} + \frac{9\sqrt{11 - \sqrt{21}}\sqrt{2}}{140}} = 0.78419$$

We currently offer a thousand dollars for a proof (or disproof) that this ratio is the best possible.

One day a letter arrived from Martin. He was writing a column on "The Steiner Problem." He had a host of examples and conjectures about more general grids. We answered, he answered, each finding new examples and counterexamples to the other's conjectures. The work was substantial enough to merit publication. This was Martin's first appearance in a refereed math journal.[33]

Our work was mostly sharp conjectures. For example, for an $n \times n$ square grid when $n = 7$, we conjectured the shortest configuration was the one shown in figure 88. We are proud to report that our joint paper won a prize for exposition. Even better, our conjectures caught the eye of a research team from Australia a few years later. They managed to rigorously prove all of our discoveries and more.[33] Martin told his own version of the story in his last *Scientific American* column, reprinted with addenda in *The Last Recreations*.[35] This all shows what we mean by saying he touches the material.

Thus, we and many others are proof: Martin has turned innocent youngsters into math professors. If you go to one of the national

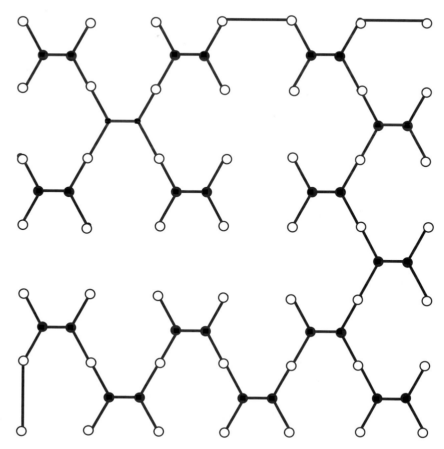

Figure 88. Shortest network
for 7 × 7 grid

mathematics meetings and ask randomly about Martin Gardner, you will find hundreds of examples of the converse, professional mathematicians who have become intrigued by one of Martin's columns and who have spent hours lost in the wonder that an interesting problem creates.

How did he do it? Perhaps Martin's greatest trick is this. He wrote mathematics so that both youngsters *and* professional mathematicians couldn't wait to see his next column. He was not a mathematician. He had an undergraduate degree in philosophy from the University of Chicago. Indeed, while he understood the idea of limits, Martin didn't really know calculus (for example, integrating $\sin^3(x)$). This is probably all to the good in communicating with nonmathematicians. How he kept professionals interested is still a mystery. One thing we noticed: Martin's writing is chock full of examples, facts, related anecdotes, and mathematical facts that tumble out with very little philosophy or filler. Also, as a nonprofessional, Martin allowed himself to be engagingly enthusiastic about his topics.

But there is more. It is not fashionable to write about talent in these days of equality. Yet we acknowledge talent in athletics, drawing, and singing. Well, Martin was as talented a writer about recreational mathematics as we are ever likely to see. It doesn't come easily. Martin told us that during his twenty-five years writing for *Scientific American,* he spent more than twenty-five days a month working full time on his monthly column. This included writing answers to countless letters and inquiries. For dealing with cranks, he had a postcard printed with a checklist of crank topics:

I cannot answer your letter about:
☐ trisecting the angle;
☐ the four color problem;
☐ doubling the cube;

. . .

because I am not trained as a professional mathematician.

And he would place a ✓ at the appropriate topic.

His network of correspondents brought brilliant recreational gems out of the woodwork. There is strength in numbers. The best of thousands of cute contributions can be profound. The great computer scientist Don Knuth has spent weeks going through Martin's files, panning for (and finding) lost gold. By Knuth's efforts, and the hard work of Stan Isaacs, Martin's files, including a folder or three for every column, his research materials, drafts, and the letters from readers, are available for qualified scholars at Stanford University's Green Library.

Here is a story we find sad. One day (around 1970) Martin seemed in unusually high spirits. "You know, after college I wrote a novel about the intricacies of modern religion in academia. I sent it around and around, and its been on the shelf all these years. I was at a party and a young publisher asked if I had anything sitting around. I've dusted off my novel and he likes it." For a year thereafter, it was all Martin could talk about. "I've finished the galleys on the novel." They've designed a great cover." "I hear it might be reviewed in the daily *New York Times*." And then, one of the saddest sentences we have ever heard: "If this thing takes off, maybe I can finally stop writing about puzzles." We found it shocking that the best writer of popular mathematics in the world wasn't deeply satisfied. We have taken our shock to heart in our own work. Each of us has talents. We vowed to enjoy our success for these and each will throw a pie at the other if he has high-blown fantasies.

Martin spent a lifetime in magic. He saw Houdini and Thurston as a lad, and every serious magician since, over a more than eighty-year period. Similarly, for the past fifty years, he had been in the absolute center of mathematical magic. We regularly called on his recollections to explain one of our omissions. An early popular book on mathematical recreations is Royal V. Heath's *Mathemagic: Magic, Puzzles and Games with Numbers*.[36] Heath performed and wrote actively. Why isn't he on our list? The answer is simple. Heath was a rich stockbroker who carried himself as a pompous fool. He performed tedious mathematical tricks that put his audience to sleep. Martin recalled Heath doing a stage performance at New York's Society of American Magicians. He covered a large blackboard with things like magic squares made up of only prime numbers. It is one thing to do this at your kitchen table and *maybe* a great showman could make it come to life, but Heath wasn't a showman and the audience was dying. Near the end of his performance, just before he delivered his closing lines, an elderly gentleman arose and said he'd like to try something different. He erased the board and had ten people call out various ten-digit numbers. As fast as they were called, he wrote them down, drew a line underneath, and instantly wrote down a giant total. He quickly erased the numbers (so no one could check) and took a deep bow. The old timer was Al

Baker, dean of American Magicians at the time. He was well-known for poking the air out of inflated egos. The audience loved it and Heath stormed off furious.

Martin never had a bad word for anyone. To our tale above he added that Heath is the reason he *had* a career writing about recreational math. Heath had a large cloth model of a fascinating "hexaflexagon" in his high-rise Manhattan apartment. Martin became intrigued and went out to Princeton to interview the four scientists (Richard Feynman, John Tukey, Arthur Stone, and Bryant Tuckerman) who had cooked up the amazing gadget. He sold the story about it to *Scientific American*. The publisher at that time, Gerard Piel, liked it and asked for more. The readers liked these and the column was born.

Figure 89. A paper hexaflexagon (image courtesy of Robert Lang)

At the time Martin was just barely eking out a living writing for children's magazines. Dai Vernon told us that Martin was really down. When the magicians met, Martin couldn't afford a meal and settled for a cup of coffee. "His cuffs were frayed and he was close to poverty for several years." He turned down corporate jobs. He wanted to make it by writing his way. He never sold out.

It's time for a bit of Martin's magic. By far, his most famous trick is his Lie Speller. Here, someone chooses a card and the name of the card is spelled out (one card for each letter) but the spectator is allowed to lie along the way. At the end, the last card dealt is the actual card that was chosen. This was a new plot (and method) that has intrigued magicians. It was first published in the American magic magazine the *Jinx*.[37] The persistent reader can find it, along with many variations, in *Martin Gardner Presents*. We asked Martin which of his tricks he likes best and he told us about an amazing geometrical vanish that he published without taking credit for his invention. It consists of a square, cut into pieces. When the pieces are slid about and rearranged, the square appears with a large interior hole (figure 90). Martin had written on extensions of geometrical vanishing puzzles (in *Mathematics, Magic and Mystery*). This version results in a much larger hole than any previous one. Martin tells this story in his own words in an interview in the *College Mathematics Journal*.[38]

This interview also contains many of Martin's favorites, including his favorite toothpick puzzle: "Move just one toothpick in the Giraffe shown [in figure 91] and leave the animal *exactly* as before except with a different orientation on the plane." (See figure 91.) The reader can find much more about these intriguing vanishing puzzles in *Games*

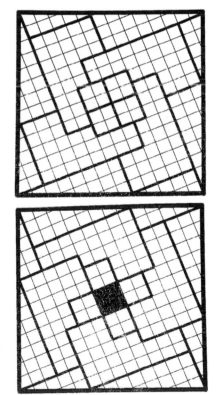

Figure 90. Martin Gardner's vanishing square

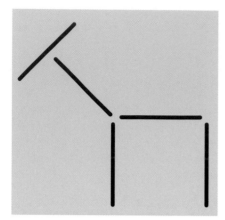

Figure 91. Martin Gardner's giraffe toothpick puzzle

Magazine and the beautiful book *Puzzles Old and New,* by Slocum and Boterman.[39]

Here is a version of a small trick of Martin's that we find charming. Find a spectator who has three initials. (We are P.W.D. and R.L.G.; Martin had no middle initial.) Say the initials are S, P, and H. You patter as follows: "There are curious coincidences all around us. I'd like to show you something strange about these initials." Write them out three times on a piece of paper and tear the paper into nine pieces.

Put one set of initials onto the table. Hand the remaining two sets to two of the spectators. "I'd like you both to help with a coincidence. Choose, in any way you like, to put your set of initials face-down below the set on the table. Just so that we don't have anything obvious, please don't put an 'S' under the 'S,' a 'P' under the 'P,' or an 'H' under the 'H'" (see figure 92). The spectators comply while you engage them in conversation ("You two didn't collude, did you?" "How long have you known S.P.H.?" etc.). To conclude, you show that an amazing coincidence has indeed occurred.

One of two things can happen. When the spectators turn up the face-down papers below each of the three face-up letters, the first possibility is that all three pairs match, i.e., the two 'S's are together, as are the two 'P's and two 'H's. If so, the face-up 'S,' 'P,' and 'H' are swept away and the coincidence is revealed (with some fanfare). Alternatively, *each* pile of three has a complete set of initials 'S,' 'P,' and 'H' (see figure 93). You conclude, "Despite your randomizing things, like seeks like and the unity of your name shines through."

The trick works itself. Martin invented it with playing cards and published it as Ace—Two—Three.[40] A marketed version with the initials E—S—P also appeared. The version with initials was shown to us by

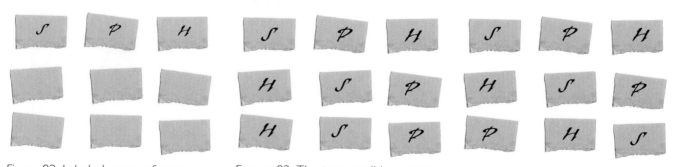

Figure 92. Labeled scraps of paper　　　　Fogure 93. The two possible outcomes

Figure 94. The authors with Martin Gardner in 2007

Frank Quinn forty years ago. Frank is long gone but his charming variation will live on.

Perhaps Martin's most serious contribution is creating the *field* of modern recreational mathematics. We owe this book and much else to this star of mathematical magic. Both of us had many pats on the back during our careers. One of the honors we are proudest of: Each of us has had Martin dedicate a book to us. It doesn't get any better than that!

Chapter 11

GOING FURTHER

Suppose you've gotten this far and want some more. More recreational mathemagic, more math, more magic. This chapter gives pointers to resources and techniques for going forward.

RECREATIONAL MATHEMATICS. It's always nice when questions have easy answers. How does one learn more recreational mathematics? What are the best, most interesting sources? Answer: Go get any (or all) of Martin Gardner's collections of *Scientific American* columns or his *Colossal Book of Mathematics* for a good sampling. If you are interested in magic, all of these books have a chapter focused there as well. The books have so much more and are so engagingly written that we bet you will get lost for hours.

A word of caution: Martin has well over 140 books in print. The most relevant ones are these sixteen:

0. *Mathematics, Magic and Mystery*
1. *Hexaflexagons and Other Mathematical Diversions*
2. *The Second "Scientific American" Book of Mathematical Puzzles and Diversions*
3. *New Mathematical Diversions*
4. *The Unexpected Hanging and Other Mathematical Diversions*
5. *Martin Gardner's Sixth Book of Mathematical Diversions from "Scientific American"*
6. *Mathematical Carnival*

7. *Mathematical Magic Show*
8. *Mathematical Circus*
9. *The Magic Numbers of Dr. Matrix*
10. *Wheels, Life and Other Mathematical Amusements*
11. *Knotted Doughnuts and Other Mathematical Entertainments*
12. *Time Travel and Other Mathematical Bewilderments*
13. *Penrose Tiles to Trapdoor Ciphers*
14. *Fractal Music, Hypercards and More Mathematical Recreations from "Scientific American"*
15. *The Last Recreations: Hydras, Eggs and Other Mathematical Mystifications*

The complete collection, without item (0), is available on one searchable CD from the Mathematical Association of America. The books are being reissued by Cambridge University Press in updated versions: Martin learned to use the Web (at ninety-five years of age!) and helped to bring the books up to date.

Item (0) is Martin's original mathematical magic book. It is still the best single (short) source. The remaining books are collections of his *Scientific American* columns enriched with extensive addenda based on the collected wisdom of Martin's many readers.

We have taught courses on mathematics and magic tricks at both Harvard and Stanford. The students are a mix of math kids and magic kids with a scattering of the curious. We try to pair the math kids with the magic kids—they really do have something to learn from each other. The first assignment is always this: Go get one of Martin Gardner's books out of the library. Find a chapter with a magic trick. Next week you will have to perform your trick and explain how it works. We recommend this homework assignment to our readers.

LEARNING MORE MATH. There has recently been an explosion of popular mathematics books. These aim at giving a flavor of the excitement going on in the world of mathematics. There are at least four books about zero, books dedicated to various basic mathematical constants ($\pi, e, i, \phi, \gamma, \sqrt{-15}, \dots$), books on focused topics such as the heroic classification of finite simple groups, the recent solution to the one million–dollar Poincaré conjecture, and at least five books on the celebrated Riemann hypothesis.

Most of these books "give a flavor," rather than attempting to clearly teach any serious mathematics. One remarkable exception, which we recommend, is Richard Courant's and Herbert Robbins's wonderful classic, *What Is Mathematics?* (1941). For a gentler introduction, we suggest Barry Mazur's *Imagining Numbers (Particularly the Square Root of Minus Fifteen)* (2003).

Higher mathematics is usually thought of as starting with calculus. Curiously, we know of essentially *no* magic tricks that lean on calculus. If the reader wants an introduction to calculus, perhaps the best is Silvanus P. Thompson's *Calculus Made Easy* (2008). It has been wonderfully updated by—Martin Gardner.

The kind of mathematics used in our book is of three basic flavors: combinatorics, number theory, and group theory. There are accessible introductory accounts of these topics in Sherman K. Stein's *Mathematics: The Man-Made Universe* (1998) and, at a slightly higher level, Herstein's and Kaplansky's *Matters Mathematical* (1978). If these appeal, there are accessible introductory courses on the Web and at many colleges through their extension programs.

LEARNING MORE MAGIC. The kind of self-working magic explained in this book is only a tiny part of the subject. There are sleight-of-hand tricks, tricks with special props and gadgets, grand illusions, mental magic, tricks aimed at kids, religious magic, bizarre magic, and so on.

To go further, we suggest a little basic sleight of hand with cards. Jean Hugard's and Frederick Braue's *Royal Road to Card Magic* (1951) is the best introduction. Further, a wonderful series of books by the Swiss magician Roberto Giobbi take the reader from self-working tricks to advanced sleight of hand. Indeed, Giobbi's multivolume *Card College* makes for the equivalent of a university degree in card magic.[1]

There is a huge magic literature aimed at serious amateurs and professionals. One of us has (about) five thousand pamphlets, books, and volumes of magic journals in our personal library. We currently subscribe to some thirty magic journals, most of these (almost surely) too technical for an outsider to get much from. There is also a large, growing collection of teaching CDs that are more easily accessible. One can find almost any trick demonstrated on YouTube and many tricks explained in Wikipedia.

How to touch this world? One route is this: Many sizable cities have a magic store and most sizable towns have magic clubs. The Society of American Magicians (S.A.M.) and the International Brotherhood of Magicians (I.B.M.) are backbones for many of these in the United States, with other countries having parallel structures.

Find a nearby magic store, drop in, and buy a book or CD. They will try to sell you a load of junk and this should be avoided until you know what you are doing. If you go on a Saturday afternoon, a bunch of magicians will be hanging around. Introduce yourself, saying you are just starting. Most likely, they will show you some tricks and tell some stories. The proprietor will be able to tell you about local magic clubs. We have found magicians to be a gentle, sociable bunch. Most magic clubs have occasional evenings aimed at the public. If you go and meet a few people, they will have advice, help correct your fledgling technique, and point you to local resources.

Like mathematics, magic has a deeper side. In addition to the secrets and the sleights, how the gadgets work and the like, there are the real secrets of magic: Presentation, misdirection, and the philosophy of magic. These make the difference between a kid who does tricks and a performance that really fools and moves people. It is difficult to lay out a road map here. If you want something to aim for, try to gear up to read, digest, and understand the series of books about the magic of Arturo Ascanio and Juan Tamariz.[2]

LEARNING MORE JUGGLING. There is a large variety of sources from which to advance your skill in juggling. The main source of all kinds of information on juggling is the Web site www.juggling.org. It carries rather complete lists of juggling videos, juggling props, juggling festivals, Web sites of many current prominent jugglers, books, meeting times and locations of local juggling clubs, etc. Many colleges have juggling clubs that welcome jugglers at all levels, especially beginners. A good beginning book is *The Art of Juggling* by Ken Benge.[3]

AND BACK. In this book, we have traveled the road from magic to mathematics and back to magic. The road doesn't stop there. We have found the back-and-forth keeps going. If you manage to use mathematical understanding to invent a good new trick, you may find that natural variations of the trick lead to new math problems. So it goes.

Two things that are hard to explain: What makes a magic trick a good trick? What distinguishes a self-working trick from a real piece of mathematics? Time and again we have seen hackneyed presentations of simple arithmetic presented as original contributions. Most self-working magic has this flavor: A poor trick devoid of mathematics. It reminds us of a "horse-cow" (i.e., it gives milk like a horse and runs like a cow). We have tried to show there is something left. There is certainly a lot left to do.

Chapter 12
ON SECRETS

A mathematician is a conjuror who gives away his secrets.
—*John Horton Conway*

Magic gets some of its appeal from its secrets. The magician knows but doesn't tell. Some spectators find this frustrating, some find it alluring. The secrets are a central part of the story. When you are accepted into the magicians' world, you agree to keep the secrets private. Those who deviate are shunned, literally thrown out of the club and not usually accepted back. David Devant, one of the great creators and performers in the early twentieth century, "erred" in writing a magic book for the public. He was thrown out of The Magic Circle, an exclusive English magic society he helped to found (he had also been its president), and was never readmitted. He spent years being looked down on by other magicians.

Things are changing but secrets are the glue of social fabric in the close-knit mix of the magic community. From our teenage years to now, if we are in a strange city or a strange part of the world, we can snoop around and find a serious local magician (usually someone who does something else for a day job). This is done with a phone call or an email: "Do we know anybody serious in Indianapolis (or Nice or Shanghai)?" When we get into town, we call perfect strangers and announce: "I'm a magician visiting from the West Coast. X told me that you were seriously interested in magic. . . ." Invariably, we are invited over, or get to meet over coffee or a meal. Joe from down the road comes along too and a "session" of some sort transpires. This is usually just gossip and trading tales, although sometimes, some wonderful magic is shown. Magicians are happy to meet others with whom they can share secrets. They also enjoy trying to fool each other.

Mathematics used to be shrouded in secrecy. Back in 287–212 BC, Archimedes, one of the truly great mathematical creators of all time, used to tantalize other mathematicians by announcing special cases of general theorems of his own and challenging them to prove them. It was recently discovered that Archimedes had essentially created calculus two thousand years before Newton and Leibniz. He could use it to see things that others couldn't imagine. He kept his secret (called "the Method") private and, when he died, it seemed that the secret died with him. It turned out that he entrusted the secret to a friend in a long-missing manuscript that has been lost forever. A copy (of a copy) surfaced in an illuminated religious book about a hundred years ago. Centuries before, Archimedes' manuscript had been scrubbed off in order to reuse the parchment. Someone spotted the remnants of math close to two thousand years later and the manuscript was rediscovered. This amazing story is well-told in *The Archimedes Codex: How a Medieval Prayer Book Is Revealing the True Genius of Antiquity's Greatest Scientist* by Reviel Netz and William Noel. Keeping secrets has its costs.

The tradition of keeping secrets lived throughout math's history. Tartaglia, a sixteenth-century mathematician, figured out how to solve cubic equations (such as find x so that $x^3 + 10x^2 + 7x = 100$). Such problems were publicized and prizes offered. When Tartaglia visited a new town, if the local sages couldn't solve the problem, Tartaglia's reputation was established. In a moment of weakness, Tartaglia explained his method to another mathematician, Cardano. Though Cardano swore secrecy, we all know what that too often means. Cardano published the method a few years later and often gets credit for the result, even today.

Nowadays, secrecy in mathematics too often comes mainly from mathematicians trying to ensure credit for their discoveries. Hundreds of years ago, Newton and Leibniz (two great mathematicians who competed with each other) communicated their work with secret codes so each could claim credit. In the twentieth century, Andrew Wiles worked on Fermat's last theorem in secret for seven years before he went public with his discovery. In the other direction, an unusual collaborative project called Polymath was recently carried out by the Cambridge mathematician Tim Gowers. He enlisted the real-time efforts of literally dozens of mathematician around the globe to collaborate via the Internet to attack a previously unsolved (and difficult) problem in combinatorial mathematics. He argued that, with our new communication technology, many brains working in concert could be more effective than individuals each struggling on his or her own. Amazingly, this experiment actually worked and several proofs of the desired result were discovered, and then generalized to prove even more. The problem then became how to allocate the credit. Gowers's plan was to publish the final proof under the name Polymath (perhaps thanking all the participants in the acknowledgments). Is Polymath a possible new paradigm in research (both in mathematics as well as magic)? Only time will tell.

The *mathematics* of secrets has emerged as a hot topic through cryptography and so-called zero-knowledge proofs in theoretical computer science. In this last application, you want to convince someone that you know a secret without giving any information about what that secret is.

Here is an example, related to us by computer scientist Moni Naor. He was playing the popular game *Where's Waldo* (sometimes called *Where's Wally*) with his seven-year-old daughter. Here, a group of players looks at a huge picture and tries to spot the tiny figure of Waldo in a mess and tumble of people. It's a hard visual recognition task and kids are as likely to succeed as adults. This time, after a minute, Moni called out, "I got it," and his daughter called out, "Liar, liar." (Nowadays, kids just don't respect their parents the way they used to!) Thus, a problem. Moni knew a secret (where Waldo was) and wanted to convince his daughter that he knew it without telling her anything further. We will tell Moni's solution at the end of our discussion, but invite the reader to speculate for now.

In the zero-knowledge proof variation, we want to convince someone that we can prove a theorem (say, Fermat's last theorem) without giving a hint of how our proof goes. Here is an example: In chapter 3, we encountered the problem of finding a cycle through a graph that starts and ends at the same point and passes through every vertex just once. For some graphs this is easy (picture a cycle on *n* points). But it doesn't seem easy to find such a cycle in general graphs without endless trial and error (e.g., see figure 1).

Indeed, a solution to the Hamiltonian cycle problem would solve a huge list of related so-called NP-hard problems. These are a vast collection of presumably difficult computational problems for which the

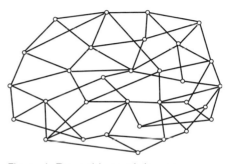

Figure 1. Does this graph have a Hamiltonian cycle?

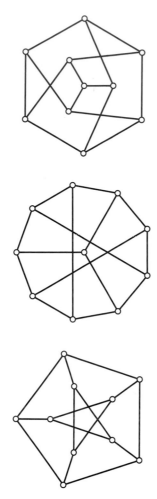

Figure 2. Three drawings of the same graph

only solutions in the general case involves a brute-force enumeration of all possibilities.[1]

Suppose we know a Hamiltonian cycle in a graph and want to convince an adversary that we know one without giving any further information. Here is the trick: The original graph is just a list of vertices and edges. We proceed by (secretly) choosing a (random) permutation of the vertices. This determines a new graph (where the old edges permute to new edges). This permutation gives an isomorphic copy of the original graph. It seems to be impossible to practically tell if two large graphs are isomorphic or not. Even when the graphs are not so large, it can be tricky. For example, in Figure 2 we show three pictures of the *same* graph drawn in three different ways.

Back to our trick, we show our new graph, write down our permutation on a piece of paper, and write down a Hamiltonian cycle in the new graph on a second piece of paper. This is easy to do because we know the permutation. Now, our adversary chooses (say, by flipping a coin) to check either our permutation, or our cycle in the new graph. In either case she learns nothing about our cycle in the original graph. Of course, doing this just once isn't convincing (we might have just made up a graph with a Hamiltonian cycle and presented it as an isomorphic graph) but by repeating this basic choice on, say, one hundred fresh instances, one must conclude that either we know a Hamiltonian cycle in the original graph or an event of probability $\frac{1}{2^{100}}$ has occurred.

The above captures the essential idea in a special case. Since a huge list of problems are known to be computationally equivalent to Hamiltonian cycle problems, the procedure is broadly useful.

What about Moni's solution? He simply did this: He took a large double-sheet of newspaper and cut out a tiny hole. He took the *Where's Waldo* book and put it under the paper, twisting and turning it so that the position of the book was random. Then he carefully positioned the hole in the paper above Waldo and showed it to his daughter without revealing anything further.

There is a curious tension for us in the keeping of secrets. We are academics. When we are around our mathematical colleagues, we are dying to tell them what we are up to. Once in a while, a nonmagician will ask us how a trick works. It feels funny to say, "Sorry, I can't tell you, I'm sworn to secrecy."

Magicians keep secrets from each other, too. Over the years, we have been privy to the inner workings of wonderful magic and we have kept these secrets. If nothing else, it reminds the magic world of the charm and allure of secrets. Juan Tamariz, the great Spanish magician, has a nice way to balance secrecy (within magic) and magic's growth as a field. When he discovers something new, he keeps it for himself for ten years, and then tells the brethren. We might point out that the tradition in juggling is just the opposite. Jugglers love to share their tricks and techniques with anyone interested in listening and learning. The feeling is that if you are willing to practice hard enough to learn some difficult juggling skill (such as spinning a three-ball high stack or juggling a cross-handed five-ball cascade), then more power to you. Of course, most of these very challenging skills don't have much commercial performance potential!

As of 2010, a profound change has taken over magic and its secrets. There has always been public exposure. A journalist will spot a method and write it up in a newspaper with fanfare. A "masked magician" will spill the beans on a television special. These are painful intrusions but they disappear within a day. What has changed magic is the Internet. Many, many tricks are now permanently on display on sites like YouTube and Wikipedia. If you see a trick and want to know how it's done, you can just type a few keywords and find out, maybe even on your iPhone during the performance.

This record of secrets exposed is permanent. It builds on itself. The same strengths that build Wikipedia, the collective wisdom of the community, conspire to take away magic's secrets. With this change, magicians themselves have become promiscuous. First in books, then on tapes, and now on DVDs and Internet chat rooms, many carefully guarded secrets are hung out for all to see. This change isn't going away.

How will it change magic? Here are three optimistic thoughts. First, with the huge sea of information, it is hard to tell the good from the bad, the right method from the wrong ones. Now the secret becomes knowing where to look, and what to believe. Maybe it's always been that way. Second, perhaps the exposure of standard methods will drive the magic community to invent new ones. A look at magic's techniques and effects reveals many a method that goes back over five hundred years. Bring on the new. We need it. Finally, maybe the exposure will allow a performance style where skill, technique, and presentation carry the day. After all, when we hear a gifted singer render a well-known song, we can still enjoy it and be moved. Dai Vernon once took us to an old-fashioned billiard parlor where three-cushion billiards was the game of choice. Instead of a rowdy group of beer drinkers, there was a silent crowd watching two masters battle it out. Quiet nods and polite applause were the crowd's response. Vernon said, "It would be wonderful if magic could be appreciated in this way."

CHAPTER 1: MATHEMATICS IN THE AIR

1. Bob Hummer, "Face-up/Face-down Mysteries" (privately printed manuscript, 1942).

CHAPTER 2: IN CYCLES

1. M. A. Martin, "A Problem in Arrangements," *Bulletin of the American Mathematical Society* 40 (1934): 859–864.
2. Charles T. Jordan, "Thirty Card Mysteries" (privately printed manuscript, 1919).
3. Sherman K. Stein, *Mathematics, the Man-Made Universe* (Mineola, NY: Dover Publications, 1999), 144–148; and "The Mathematician as an Explorer," *Scientific American* (May 1961), 149–158.
4. Hal Fredricksen, "A Survey of Full Length Nonlinear Shift Register Cycle Algorithms," *SIAM Review* 24, no. 2 (1982): 195–221; and Anthony Ralston, "De Bruijn Sequences-A Model Example of the Interaction of Discrete Mathematics and Computer Science," *Mathematics Magazine* 55, no. 3 (1982): 131–143.
5. Donald Knuth, *The Art of Computer Programming*, vol. 4A, *Combinatorial Algorithms, Part 1* (Reading, MA: Addison-Wesley Publishing Co., 2011).
6. See http://theory.cs.uvic.ca/~cos/.

CHAPTER 3: IS THIS STUFF ACTUALLY GOOD FOR ANYTHING?

1. http://en.wikipedia.org/wiki/digital_pen.
2. See Glenn H. Hurlbert, Chris J. Mitchell, and Kenneth G. Paterson, "On the Existence of de Bruijn Tori with Two by Two Windows," *Journal of Combinatorial Theory, Series A* 76, no. 2 (1996): 213–230; and F. J. MacWilliams and N.J.A. Sloane, "Pseudo-random Sequences and Arrays," *Proceedings of the IEEE* 64 (1976): 1715–1729.
3. Joshua N. Cooper and Ron Graham, "Generalized de Bruijn cycles," *Annals of Combinatorics* 8, no. 1 (2004): 13–25.
4. Leo Marks, *Between Silk and Cyanide: A Codemaker's War*, 1941–1945 (New York: Harper Collins Publishers, 1998).
5. A clear account appears in Pavel Pevzner, Haixu Tang, and Michael Waterman, "An Eulerian Path Approach to DNA Fragment Assembly," *Proceedings of the National Academy of Sciences* 98 (2001): 9748–9753. See also Zhang Yu and Michael S. Waterman, "An Eulerian Path Approach to Local Multiple

Alignment for DNA Sequences," *Proceedings of the National Academy of Sciences of the United States of America* 102, no. 5 (2005): 1285–1290.

6. For further details see Pavel A. Pevzner, "Computational Molecular Biology: An Algorithmic Approach," *Computational Molecular Biology* (Cambridge, MA: MIT PRESS, 2000); Richard M. Karp, "Mathematical Challenges from Genomics and Molecular Biology, *Notices of the American Mathematical Society* 49, no. 5 (2002): 544–553; and Richard M. Karp, Li Ming; Pavel Pevzner, Ron Shamir, "Foreword," *The Journal of Computer and System Sciences* 73, no. 7 (2007): 1023.

7. See Stephen M. Stigler, *The History of Statistics: The Measurement of Uncertainty before 1900* (Cambridge, MA: Harvard University Press, 2003).

8. Persi Diaconis and Silke W. W. Rolles, "Bayesian Analysis for Reversible Markov Chains," *Annals of Statistics* 34, no. 3 (2006): 1270–1292.

9. See http://www.cs.uvic.ca/ruskey/.

10. Robert J. Johnson, "Long Cycles in the Middle Two Layers of the Discrete Cube," *Journal of Combinatorial Theory, Series A* 105, no. 2 (2004): 255–271.

11. Visit http://www.claymath.org/millennium/ for details.

12. G. Hurlbert, B. Jackson, and B. Stevens, "Research Problems on Gray Codes and Universal Cycles," *Discrete Mathematics* 309 (2005): 5332–5340.

CHAPTER 4: UNIVERSAL CYCLES

1. Ronald L. Graham, Donald E. Knuth, and Oren Patashnik, *Concrete Mathematics. A Foundation for Computer Science*, 2nd ed. (Reading, MA: Addison-Wesley Publishing Company, 1994).

2. Robert Johnson, "Universal Cycles for Permutations," *Discrete Mathematics* 309, no. 17 (2009): 5264–5270.

3. P. Erdős, R. L. Graham, I. Z. Ruzsa, and E. G. Straus, "On the Prime Factors of $\left(\frac{2n}{n}\right)$," *Mathematics of Computation* 29 (1975), 83–92.

4. Persi Diaconis and Ron Graham, "Products of Universal Cycles," *A Lifetime of Puzzles: Honoring Martin Gardner*, ed. E. Demaine, M. Demaine, and T. Rodgers (Wellesly, MA: A K Peters, Ltd., 2008), 35–55.

CHAPTER 5: FROM THE GILBREATH PRINCIPLE TO THE MANDELBROT SET

1. Paul Curry, "Paul Curry Presents" (privately printed manuscript, 1974).

2. David Ben, *Zarrow, A Lifetime of Magic* (Fair Lawn, NJ: Meir Yedid Magic, 2008).

3. http://www.neave.com/fractal/.

4. H. O. Peitgen, H. Jürgens, and D. Saupe, Chaos and Fractals: *New Frontiers of Science* (New York: Springer, 1993).

5. D. Sullivan, "Bounds, Quadratic Differentials, and Renormalization Conjectures," *Mathematics into the Twenty-first Century*, ed. F. Browder (Providence, RI: American Mathematical Society, 1992), 417–466.

6. Tan Lei, ed., *The Mandelbrot Set: Theme and Variations* (Cambridge: Cambridge University Press, 2000).

7. A. Weiss and T. D. Rogers, "The Number of Orientation Reversing Cycles in the Quadratic Map," *Oscillations, Bifurcation, and Chaos: CMS Conference Proceedings*, vol. 8 (Providence, RI: American Mathematical Society, 1987), 703–711.

8. Jason Fulman, "Applications of Symmetric Functions to Cycles and Increasing Subsequence Structure after Shuffles," *Journal of Algebraic Combinatorics* 16 (2002): 165–194.

9. C. McMullen, "The Mandelbrot Set Is Universal," *The Mandelbrot Set: Theme and Variations.*

10. N. G. de Bruijn, "A Riffle Shuffle Card Trick and Its Relation to Quasicrystal Theory," *Nieuw Archief voor Wiskunde* (4) 5, no. 3 (1987): 285–301.

11. Marjorie Senechal, *Quasicrystals and Geometry* (Cambridge: Cambridge University Press, 1995); Charles Radin, *Miles of Tiles* (Providence, RI: American Mathematical Society, 1999); and Martin Gardner, *Penrose Tiles to Trapdoor Ciphers* (Washington, DC: The Mathematical Association of America, 1997).

12. Donald Knuth, *The Art of Computer Programming*, vol. 3, *Sorting and Searching* (Reading, MA: Addison-Wesley Publishing Co., 1973), section 5.4.1.

13. *Genii* 52 (1989), 574–577, 658–665, 743–744.

14. Justin Branch, "Cards in Confidence," vol. 1 (Plymouth, England: privately published manuscript, 1979), 100–147.

CHAPTER 6: NEAT SHUFFLES

1. Karl Fulves, *The Magic Book: A Conjurer's Collection of Dazzling Effects for Parlor and Platform* (Boston: Little, Brown and Co., 1977), 68.

2. S. Brent Morris, *Magic Tricks, Card Shuffling, and Dynamic Computer Memories* (Washington, DC: Mathematical Association of America, 1998).

3. Sarnath Ramnath and Daniel Scully, "Moving Card i to Position j with Perfect Shuffles," *Mathematics Magazine* 69, no. 5 (1996): 361–365.

4. Persi Diaconis and Ron Graham, "The Solutions to Elmsley's Problem," *Mathematical Horizons* 14 (2007): 22–27.

5. *Whole Art and Mystery of Modern Gaming* (London: J. Roberts and T. Cox, 1726). This has recently been republished and is available on the Amazon.com Web site, among other places.

6. I. N. Herstein and I. Kaplansky, *Matters Mathematical* (New York: Chelsea Publishing Co., 1978).

7. For references and many further details, see Persi Diaconis, Ron Graham, and William Kantor, "The Mathematics of Perfect Shuffles," *Advances in Applied Mathematics* 4, no. 2 (1983): 175–196.

8. John Gale, *Gale's Cabinet of Knowledge*, 2nd ed. (London: W. Kemmish, 1797).

9. In fact, Gale copied this from W. Hopper's *Rational Recreations* (London, 1774). Hopper translated this material from Edme Guyot's *Nouvelles récréations*, vol. 2 (Paris, 1769).

CHAPTER 7: THE OLDEST MATHEMATICAL ENTERTAINMENT?

1. See Glen Gravatt, "The Collected Writings of Glen Gravatt" (privately printed manuscript), 116. See also *Genii* 27, no. 12 (1963): 518.

2. Prevost's book has finally been translated into English. J. Prevost, *Clever and Pleasant Inventions*, part 1, trans. Sharon King (Seattle: Hermetic Press, 1998).

3. See W. Kalush, "Sleight of Hand with Playing Cards prior to Scot's Discoveries," *Puzzlers' Tribute: A Feast for the Mind*, ed. D. Wolfe and T. Rodgers (Natick, MA: A. K. Peters, 2002), 119–124.

4. S. W. Clarke, *The Annals of Conjuring*, 2nd ed., ed. E. Dawes and T. Karr (Seattle: The Miracle Factory, 2001).

5. Albrecht Heeffer, "*Récréations Mathématiques* (1624): A Study on Its Authorship, Sources, and Influence," *Gibecière* 1, no. 2 (2006). See also http://logica.ugent.be/albrecht/thesis/Ettenintro.

6. Laurence E. Sigler, trans., *Fibonacci's Liber Abaci* (New York: Springer, 2002).

7. See Oystein Ore, *Cardano, the Gambling Scholar* (Mineola, NY: Dover Publications, 1965), 132.

8. George Seber, "Capture-Recapture Methods," *Encyclopedia of Statistical Sciences*, vol. 1, ed. Samuel Kotz, Normal L. Johnson, and Campbell B. Read (New York: John Wiley and Sons, 1982).

CHAPTER 8: MAGIC IN THE BOOK OF CHANGES

1. Richard Wilhelm and Cary F. Baynes, *The I Ching or Book of Changes* (Princeton, NJ: Princeton University Press, 1950).

2. See, for example, Kidder Smith Jr., Peter K. Bol, Joseph A. Adler, and Don J. Wyatt, *Sung Dynasty Uses of the "I Ching"* (Princeton, NJ: Princeton University Press, 1990).

3. James Franklin, *The Science of Conjecture: Evidence and Probability before Pascal* (Baltimore: Johns Hopkins University Press, 2002).

4. Mark Elvin, "Personal Luck: Why Premodern China—Probably—Did Not Develop Probabilistic Thinking," *Concepts of Nature: A Chinese-European Cross-Cultural Perspective*, ed. Hans-Ulrich Vogel and Günter Dux, (Leiden: Brill, 2010), 400–468.

5. Knuth, *The Art of Computer Programming*, vol. 4A, 48–49.

6. F. van der Blij, "Combinatorial Aspects of the Hexagrams in the Chinese *Book of Changes*," *Scripta Math.* 28 (1967): 37–49.

7. Knuth, *The Art of Computer Programming*, vol. 4A.

8. The name of the trick in Chinese characters is 蘇武牧羊.

9. The book's title in Chinese characters is 中外戲法圖説. It was published in 1889 but we don't have any further information.

10. Persi Diaconis and Charles Stein, "Tauberian Theorems Related to Coin Tossing," *Annals of Probability* 6 (1978): 483–490. This gives references to related literature.

CHAPTER 9: WHAT GOES UP MUST COME DOWN

1. The proof of the correctness of this formula is not completely trivial and was first discovered in Joe Buhler, David Eisenbud, Ron Graham, and Colin Wright, "Juggling Drops and Descents," *American Mathematical Monthly* 101, no. 6 (1994): 507–519.

2. For example, see Richard Ehrenborg and Margaret Readdy, "Juggling and Applications to *q*-analogues," *Discrete Mathathematics* 157, nos. 1–3 (1996): 107–125; Chung Fan, Anders Claesson, Mark Dukes, and Ronald Graham, "Descent Polynomials for Permutations with Bounded Drop Size. *European Journal of Combinatorics* 31, no. 7 (2010): 1853–1867; or Anthony Mays's nice paper, "Combinatorial Aspects of Juggling," available on his Web site at www.ms.unimelb .edu.au/personnel/postgraduate.php.

3. Currently available at http://www.jongl.de/. Another very nice Web site is http://jugglinglab.sourceforge.net/bin/example_gen.html.

4. The Juggling Information Service Web site: www.juggling.org.

5. Ibid.

6. Ken Benge, *The Art of Juggling* (New York: Brian Dube, Inc., 1984).

CHAPTER 10: STARS OF MATHEMATICAL MAGIC

1. In this section, "we" refers more specifically to your first author.

2. See T. Kreisel and S. Kurz, "There Are Integral Heptagons, No Three Points on a Line, No Four on a Circle," *Discrete and Computational Geometry* 39 (2008): 786–790.

3. For references and further discussion, see http://garden.irmacs.sfu.ca/ or B. Mazur, "On the Passage from Local to Global in Number Theory," *Bulletin of the American Mathematical Society* 29 (1993): 14–50.

4. Stephen Minch, *Collected Works of Alex Elmsley*, vols. 1 and 2 (Tahoma, CA: L&L Publishing, 1991, 1994).

5. Ibid, vol. 1, 313–324.

6. Lisa Marburg Goodman, ed., *Death and the Creative Life: Conversations with Prominent Artists and Scientists* (New York: Springer, 1981).

7. Robert Harbin, *Paper Magic: The Art of Paper Folding* (Boston: Charles T. Brandford Company, 1957), 29.

8. Robert Neale, *Folding Money Fooling: How to Make Entertaining Novelties From Dollar Bills* (Washington, DC: Kaufman and Company, 1997).

9. Robert Neale, *This Is Not a Book* (Seattle: Hermetic Press, 2008).

10. Carsten Thomassen, "The Jordan-Schönflies Theorem and the Classification of Surfaces," *American Mathematical Monthly* 99, no. 2 (1992): 116–130.

11. Nick Trost, "Expert Gambling Tricks" (privately printed manuscript, 1975).

12. See Björner, Lovász, and Shor, "Chip-firing Games on Graphs," *European Journal of Combinatorics* 12, no. 4 (1991): 283–291.

13. Graham, Knuth, and Patashnik, *Concrete Mathematics: A Foundation for Computer Science*, 299–301; and Persi Diaconis, Ron Graham, and Suan Holmes, "Statistical Problems Involving Permutations with Restricted Positions, State of the Art in Probability and Statistics," *Institute of Mathematical Statistics Lecture Notes—Monograph* Series 36 (1999): 195–222.

14. See M. Davis, "Hilbert's Tenth Problem Is Unsolvable," *American Mathematical Monthly* 80 (1973): 233–269.

15. Martin Gardner, *Mathematical Circus* (Washington, DC: The Mathematical Association of America, 1996), chapter 13.

16. Allan Slaight, *The James File*, vols. 1–2 (Seattle: Hermetic Press, 2000), chapter 14.

17. See http://www.maa.org/columns/colm/cardcolm200706.html.

18. See, for example, J. D. Fulton and W. L. Morris, "On Arithmetical Functions Related to the Fibonacci Numbers," *Acta Arithmetica* 16 (1969): 105–110.

19. John R. Burnham, *The Contest Story* (Philadelphia: Dorrance and Co., 1951), 115.

20. Harry Riser, "The Riser Repertoire," *Magic Unity Might* 74, no. 12 (1985): 36.

21. Burnham, *The Contest Story*, 116.

22. Charles Jordan, "Psychola," *Sphinx* (February 1915), 245.

23. See D. Bayer and P. Diaconis, "Trailing the Dovetail Shuffle to Its Lair," *Annals of Applied Probability* 2 (1992): 294–313.

24. For a friendly treatment see Brad Mann, "How Many Times Should You Shuffle a Deck of Cards?" *Topics in Contemporary Probability and Its Applications*, ed. J. Laurie Snell (Boca Raton, FL: CRC Press, 1995), 261–289.

25. For an advanced survey of where Jordan's trick leads, see Persi Diaconis, "Mathematical Developments from the Analysis of Riffle Shuffling," *Groups, Combinatorics, and Geometry*, ed. A. A. Ivanov, M. W. Liebeck, and J. Saxl (River Edge, NJ: World Scientific, 2003), 73–97. For recent developments, see S. Assaf, P. Diaconis, and K. Soundararajan, "A Rule of Thumb for Riffle Shuffling," *Annals of Applied Probability* (forthcoming). This shows how fewer shuffles suffice if only some features are of interest (e.g., values but not suits). Mark Conger shows, in "A Better Way to Deal the Cards," *American Mathematical Monthly* 117 (2010): 686–700, how different ways of dealing can affect things. There is even a connection between shuffling and the "carries" process that occurs when we add numbers. See P. Diaconis and J. Fulman, "Carries, Shuffling and an Amazing Matrix," *American Mathematical Monthly* 116 (2009): 788–803.

26. Slaight, *The James File*. See also H. Lyons and A. Slaight, eds., *Stewart James in Print* (Toronto: Jogestja Ltd., 1989).

27. Bob Hummer, "Mental Colorama," *Genii* (November 1958), 83.

28. See Max Abrams's article in *Genii* (March 1990), 584.

29. Bob Neale, "End Game," *Pallbearers Review* 3, no. 4 (1973).

30. See Martin Gardner's column in *Scientific American* (August 1977).

31. See Martin Gardner's columns in *Scientific American* (October 1976) and *Scientific American* (January 1971).

32. See Warren D. Smith and J. MacGregor, "On the Steiner Ratio in 3-space," *Journal of Combinatorial Theory*, Series A, 69, no. 2 (1995): 301–332.

33. Martin Gardner, "Steiner Trees on a Checkerboard," *Mathematics Magazine* 62 (1989): 83–96.

34. M. Brazil, T. Cole, J. H. Rubenstein, D. A. Thomas, J. F. Weng and N. C. Wormald, "Minimal Steiner Trees for $2^k \times 2^k$ Square Lattices," *Journal of Combinatorial Theory, Series A*, 73 (1996): 91–110.

35. Martin Gardner, *The Last Recreations* (New York: Springer-Verlag, 1997), chapter 22.

36. Royal V. Heath, *Mathemagic: Magic, Puzzles and Games with Numbers* (Mineola, NY: Dover Publications, 1953).

37. Martin Gardner, "Lie Speller," *Jinx* (Winter Extra, 1937–1938).

38. Interview with Martin Gardner in *College Mathematics Journal* 40 (2009): 158–161.

39. See *Games Magazine* (November–December 1980), 14–18; Jerry Slocum and Jack Botermans, *Puzzles Old and New: How to Make and Solve Them* (Seattle: University of Washington Press, 1988), 144.

40. Martin Gardner, *Martin Gardner Presents* (Washington, DC: Kaufman and Company, 1993), 226.

CHAPTER 11: GOING FURTHER

1. Robert Giobbi, *Card College*, vols. 1–5 (Seattle: Hermetic Press, 2003).

2. Arturo de Ascanio, *The Magic of Ascanio* (Madrid: Páginas Magic Books, 2008); Jesús Etcheverry, *The Magic of Ascanio: The Structural Conception of Magic* (Madrid: Páginas Magic Books, 2005); Juan Tamariz, *Bewitched Music*, vol. 1, *Sonata* (Madrid: Editorial Frakson Magic Books, 1991).

3. Benge, *The Art of Juggling.*

CHAPTER 12: ON SECRETS

1. For more details see Michael R. Garey and David S. Johnson, *Computers and Intractability: A Guide to the Theory of NP-completeness* (San Francisco: W. H. Freeman and Co., 1979).

INDEX

AAG principle, 188
Abracadabra journal, 153
Abrams, Max, 207
Ace-Two-Three, 218–19
Adleman, Len, 38
advance man, 104
American Mathematical Society, 17
Animal-Vegetable-Mineral, 55
Annals of Conjuring, The (Clarke), 113
Annemann, Theo, 173
applications: genetics and, 38–42; making codes and, 34–38; robotic vision and, 30–34
Archimedes, 226
Archimedes Codex, The: How a Medieval Prayer Book Is Revealing the True Genius of Antiquity's Greatest Scientist (Netz and Noel), 226
Art of Computer Programming, The (Knuth), 25–26, 80, 126
Art of Dying, The (Neale), 160
Art of Juggling, The (Benge), 152, 223
audience, 1, 3–4, 47, 116; binary numbers and, 127, 132; Dead Man's Hand and, 175; de Bruijn sequences and, 17–19; Fibonacci trick and, 114; Gardner and, 211; Gilbreath Principle and, 67; Heath and, 216–17; Hummer and, 8, 14–15, 202, 206, 209; *I Ching* tricks and, 119, 131, 134–35; Johnson and, 45; Mathematical Three Card Monte and, 204–5; mind-reading computer and, 88; Miracle Divination and, 105, 112; Neale and, 161, 170, 210; robotic vision and, 33; universal cycles and, 57
Australian National University, 125
Australian shuffle, x, 85, 88–89, 91, 98–102

Baby Hummer, 4–7, 11, 13–14
Bachet, Gaspard, 107–8, 115–16
Baker, Al, 216–17
Banff International Research Station, 42–44
base ten, 26

base two: de Bruijn sequences and, 20–29; perfect shuffles and, 95. *See also* binary numbers
Bayer, D., 199–200
Bayes, Thomas, 42
Beckett, Laurie, xi, 132
Beckett, Samuel, 25
Bellis, Mark, xi
Bell Laboratories, 133, 200, 212
Bell numbers, 55–57
Ben, David, 67
Benge, Ken, 152, 223
Benjamin, Art, xi
Between Silk and Cyanide: A Codemaker's War, 1941–45 (Marks), 35–36
Bey, Mohammed, 173
Bible codes, 43
binary numbers: combs and, 33–34; cryptography and, 34–38; de Bruijn sequences and, 20–30; Elmsley and, 95, 156; Gilbreath Principle and, 71; *I Ching* and, 127, 132; making codes and, 34–38; perfect shuffles and, 93, 95, 156; robotic vision and, 30–34
blackjack, 98
Blackledge, J. Elder, 191, 193, 194f–198f
Blow Book, 116
Bossi, Vanni, 113
Botermans, Jack, 218
Box of Grain Transposition, The, 116
Branch, Justin, 83
Braue, Frederick, 222
British Museum, 158
Burnham, J., 191, 193
Burnt Thread, 116
busking, 203
Butler, Steve, xi

Cabinet of Knowledge (Gale), 102
cable channels, 34
Cage, John, 122–24

calculus, 32, 118, 215, 222, 226
Calculus Made Easy (Thompson), 222
Cambridge University, 221, 226–27
capture/recapture estimation, 115
Cardano, Gerolamo, 115, 226
Card College (Giobbi), 222
Card Colm, 188
Cardini, 173
Cardiste (Duck), 93
cards: de Bruijn sequences and, 18 (*see also* de Bruijn sequences); Gilbreath Princple and, ix, 61–62, 64, 68–72, 79–83; Kruskal principle and, 212; number of magic tricks and, 114–18; shuffles and, x, 84–102, 193 (*see also* shuffles); stay-stack arrangement and, 88–90, 94, 154
Cards in Confidence (Branch), 83
CATO ("cut and turn over two") shuffle, 7
chain moves. *See* Inside-Outside
Chaos and Fractals: New Frontiers of Science (Peitgen, Jürgens, and Saupe), 77
CHaSeD order, 54
chat rooms, 229
cheating trick, 66–67
Chop-Chop's Mental Colorama, 205–6
Christ, Evelyn Pilliner, 173
Christ, Henry, 200; background of, 173; Dead Man's Hand and, 175–77; inner circle of magicians and, 173, 177; *Jinx* magazine and, 173, 175; roulette system and, 177–81; sleight-of-hand and, 173, 177
Christ, Michael, xi, 173
Christ, Richard, 173
Christ, Robert, 173
Clarke, S. W., 113
close-up tricks, 154
codes: Bible, 43; computers and, 35, 37–38; corrupting strings and, 35–36; genetic, 38–42; Gray, 25–26, 42–43; making, 30, 34–38, 211–12, 227. *See also* cryptography
Collected Works of Alex Elmsley (Minch), 159

College Mathematics Journal, 217–18

Colossal Book of Mathematics (Gardner), 220

Coluria, 25, 54–55, 190

combinatorics: de Bruijn sequences and, 17–29, 42–46; employment from, 42–46; Gilbreath Principle and, 67; going further with, 222; Gray Codes and, 25–26, 42–43; Hamiltonian cycles and, 43–46; hexagrams and, 121–27, 134; *I Ching* and, 121, 126; Internet collaboration and, 227; juggling and, 138–44; McKay and, 42–43; Mills Mess and, 144; Polymath project and, 226–27; practical use of, 42–46; robotic vision and, 32; universal cycles and, 47–60

combs, 33–34

computers, 28, 153, 227; codes and, 35, 37–38; de Bruijn sequences and, 23–25, 45; designing sorting algorithms and, 79–80; Elmsley and, 156–58; fingers as, 204–5; Game of Life and, 211; Gilbreath Principle and, 63–65; hackers and 34, 38; Hummer and, 204–5, 208; *I Ching* and, 126, 133; Jordan Curve Theorem and, 167; juggling and, 144; Knuth and, 25–26, 80, 126, 215–16; Mandlebrot Set and, 72; mind-reading trick and, 85–92; shuffles and, 85–96; sorting algorithms and, 79–80; superblock striping technique and, 80; universal cycles and, 48–49, 55, 59–60

Connely, Robert, 181

Conway, John Horton, 101, 211, 225

Cooper, Joshua N., 34

corrupting strings, 35–36

Courant, Richard, 222

cryptography, 30, 227; as big business, 38; binary numbers and, 34–38; cable channels and, 34; corrupting string and, 35–36; Data Encryption Standard (DES) and, 37–38; de Bruijn sequences and, 36–38; export licenses and, 37–38; Gardner and, 211; hackers and, 34, 38; Institute for Defense Analyses and, 212; machines for, 37–38; modulo arithmetic and, 34–36; music files and, 34; National Security Agency and, 38; public-key, 211; RSA and, 38; S-boxes and, 37–38; TV shows and, 37–38

cubic equations, 226

Cunningham, Merce, 123

Curry, Paul, 65

Cut-Off Nose, 116

cycle structures, 78, 102

Data Encryption Standard (DES), 37–38

Dazzling Mills Family, 144

Dead Man's Hand, 175–77

Death and the Creative Life: Conversations with Prominent Artists and Scientists (Goodman), 160

de Bruijn, N. G., ix, 79

de Bruijn arrays, 31–34

de Bruijn sequences, 39, 189; audience and, 18; Banff International Research Station and, 42–44; binary numbers and, 20–30; combinatorics and, 17–29, 42–46; combs and, 33–34; cryptography and, 36–38; de Finetti's theorem and, 41–42; DNA and, 40–42; effect and, 18; elegant solution for, 26–29; employment from, 42–46; Euler and, 20–21, 43; Fredricksen and, 43; generalized, 43, 107; going further with, 25–26; graph theory and, 17, 20–25; greedy algorithm and, 23–24; Hamiltonian cycles and, 43–46; Hurlbert and, 43; Johnson and, 43–45; Knuth and, 25–26; Markov chains and, 41–42; middle-layer problem and, 43–45; modulo arithmetic and, 36; Moreno and, 43; "order matters" study and, 47–52; practical applications of, 17, 42–46; robotic vision and, 30–34; S-boxes and, 37–38; secrets and, 18–24; universal cycles and, 17–26, 43, 47–52; window length and, 19–20, 24, 30–36, 49; without lists, 23, 26–29

de Bruijn tori, 32

de Finetti's theorem, 41–42

dePaulis, Hjalmar, 113

dePaulis, Thierry, 113

Devant, David, 225

Diaconis, Persi, ix–x, 25, 133, 158, 200

digital pens, 31

Discoverie of Witchcraft (Scot), 106–7, 113–16

Discrete Mathematics journal, 46

divination tricks: Fibonacci and, 113–14; four-object, 107–12; *I Ching* and, 120, 122–25; Miracle Divination and, 105–14; modulo arithmetic and, 110–11; three-object, 107, 112–14

DNA, 17, 30, 38, 47; de Bruijn sequences and, 40–42; de Finetti's theorem and, 41–42; Euler's theorem and, 39; graph theory and, 39–42; protein folding and, 42

dovetail shuffle, 193, 199–200

down-and-under shuffle, x, 85, 88–89, 91, 98–102

Drillenger, Harry, 103–4

Duck, J. Russell, 93–94

Eating a Knife, 116

Eight Kings, 54

Elements of the Differential and Integral Calculus (Granville, Smith, and Longley), 118

Elmsley, Alex: Animal-Vegetable-Mineral and, 55; computer tricks and, 156–58; Ghost Count and, 156; irrational numbers and, 159; mind-reading tricks and, 90; museum trick of, 156–57; Penelope's Principle and, 159–160; perfect shuffles and, 92–95, 156; sleight-of-hand and, 156, 159

Elmsley Count, 156

Elvin, Mark, 125

Encyclopedia of Statistics (Seber), 116

Endfield, Cy, 208–11

Endfield, Maureen, 208

End Game, 210

endless chain. *See* Inside-Outside

Entertaining Card Magic (Endfield), 208

espionage, 17, 34–35, 38, 47

Euler, Leonhard, 20–21, 39–40, 43, 143

Euler algorithm, 40

Eulerian cycles, 43–44

Eulerian number identities, 143

Euler's theorem, 39

exchangeability, 41–42

export licenses, 37–38

"Face-up/Face-down Mysteries" (Hummer), 4

factorials, 6, 49–50, 68

faro shuffle, x, 98

Fendel, Joe, xi, 184–85

Fermat, Pierre de 125

Fermat's last theorem, 226

Ferrell, Jerry, xi

Fibonacci, 113–14

"Fibonacciana" (Gardner), 187

Fibonacci numbers, 187–89

Fibonacci Quarterly, The (journal), 187

Findley, 173

five-ball cascades, 143, 229

Five Card Mental Force, 158

Flynn, Dennis, 169

four-object trick, 107–12

fractals, ix, 61, 72–83

Franklin, James, 125

Fredricksen, Hal, 24–25, 43

Freeman, Steve, xi, 7–8, 11, 13, 15

Fulman, Jason, 78

Fulves, Karl, 25, 92

Gale, John, 102

gambling; endless chain and, 162, 166–69; probability and, 125; roulette system and, 177–81; shuffles and, 84, 102

Game of Life, 211

Games Magazine, 217–18
Garcia, Frank, 156
Gardner, Martin, ix–x; Ace-Two-Three and,
218–19; Christ and, 177; communication
skills of, 215; correspondents of, 215–16;
cryptography and, 211; as debunker of
occult and, 211; Fibonacci numbers and,
187; files of, 215–16; first research paper
of, 212–14; Game of Life and, 211; Giraffe
puzzle and, 217–18; hexaflexagons and,
217, 220; Hummer and, 207; influence of,
211, 214–21; James and, 187; Knuth and,
215–16; lack of mathematical training of,
211–12, 215; letter of recommendation
of, 211–12; Lie Speller and, 217; mathematical magic and, 216–19; Penrose tiles
and, 79, 211; popular science and, 211;
recreational mathematics and, 211, 215,
217, 219–21; roulette system and, 178–79;
Scientific American column of, 82, 211,
214–15, 217; sleight of hand and, 211;
Steiner point and, 213–14; three-initials
trick and, 218–19; Vernon and, 217; writings of, 177–78, 211–17, 220–22
Geiger counters, 35
genetics. *See* DNA
Genii magazine, 82, 206
Gentleman Joe Palooka (film), 208
geometry: Gardner and, 217; Mandelbrot set
and, ix, 61, 72–78; Monge and, 96; Penrose
tiles and, ix, 79, 211; proof trends and,
167; Say Red and, 181; summing of series
and, 179; two-dimensional, 33; vanishing
square and, 217
Ghost count, 156
Gilbert, Ed, 200
Gilbreath, Norman, ix, 62–66, 82
Gilbreath Principle: binary numbers and, 71;
cheating trick and, 66–67; combinatorics
and, 67; concept of, 61–67; de Bruijn and,
79; Endfield and, 209; First, 62–65, 71, 79,
82; Gardner and, 82; Hummer and, 207;
Knuth and, 80; lying trick and, 63–65;
Mandelbrot set and, ix, 61, 72–78; Möbius
function and, 78; modulo arithmetic and,
69–72; Penrose tiles and, ix, 79, 211; permutations and, 68–71, 75–80; red/black
trick and, 61–66; Second, 63, 65–66, 82;
shuffles and, 61–72, 79–83, 211; Ultimate,
61, 63, 68–72, 79–80, 83
"Gilbreath Principles" (Muller), 83
Giobbi, Roberto, 222
Giraffe puzzle, 217–18
Goodman, Lisa Marburg, 160

Gowers, Tim, 226–27
Graham, Ché, xii
Graham, Fan Chung, xii, 50, 57, 213
Graham, Ron, ix–x, 34, 108–12, 133
Grandmother's Necklace, 116
graph theory: de Bruijn sequences and,
17–25, 30–36, 49; de Finetti's theorem
and, 41–42; DNA and, 40–42; edges and,
20; Euler and, 20–21; Hamiltonian cycles
and, 42–46, 227–28; loops and, 20; Steiner
point and, 212–14; vertices and, 20, 45
Gray Codes, 25–26, 42–43
Greater Magic (Hilliard), 104–5, 168
greedy algorithm, 23–24
Green Library (Stanford), 216
group theory, 222
G-scam deal, 134–36

hackers, 34, 38
Hales, Thomas, 167
Hamiltonian cycles, 50; de Bruijn sequences
and, 43–46; Eulerian cycles and, 43;
importance of, 43–44; Johnson and, 43–45;
middle-layer problem and, 43–45; million-
dollar prize for, 44; nonstarting vertex and,
45; NP-hard problems and, 227–28
Han Dynasty, 130
Harbin, Robert, 162
Harry Ells Junior High School, 118
Harvard University, 32, 119, 184, 212, 221
Harvard-Yenching Institute, 119
Heath, Royal V., 216–17
Heeffer, Albrecht, xi, 113
Herpick, Jimmy, 156
Herstein, I. N., 99, 222
Hesse, Hermann, 125
hexaflexagons, 217, 220
hexagrams, 121–27, 134
Hickock, Wild Bill, 175
Hilbert, David, 187
Hilliard, John Northern, 104–5
Holmes, Susan, xii
Horowitz, Sam, 173
Houdini, 216
Howard Thurston's Magical Extravaganza, 104
Hubert's Flea Museum, 17
Hudson, Charles, 4, 13, 82
Hugard, Jean, 222
Hummer, Bob, x, 134, 162; background of,
202–4; Gardner and, 207; as genius, 201–2;
Le Paul and, 202; Mathematical Three
Card Monte and, 202, 204–8; Miraskill
and, 186; Rock, Paper, Scissors and,
210–11; Royal Hummer trick and, 8–15;

sleight of hand and, 202; ten-card trick of,
4–8; "turn two and cut at random" shuffle
and, 4–7, 11, 14; Voodoo Fortune Telling
and, 186
Hurlbert, Glenn, 43, 45, 59

IBM punch cards, 154
I Ching (Book of Changes), x, 158; authors'
talk on, 119–20; background of, 121–22;
combinatorics and, 121, 126; divination
and, 122–25; probability and, 119–20,
125–27, 134, 136; tricks based on, 127–36;
uncertainty and, 125
Imagining Numbers (Mazur), 222
impossible objects, 161
in-shuffles, 92–95, 101, 156
Inside-Outside: cheating throws and, 168; fair
throws and, 167–68; Jordan Curve Theorem and, 167; Judah and, 168–69; Neale
and, 164–73; performance of, 164–67,
170–71; secrets of, 167–71; superfair
throws and, 168–70; Vernon and, 171, 173
Institute for Defense Analyses, 212
International Brotherhood of Magicians
(IBM), 82, 223
International Jugglers' Association, 144
Internet, 227, 229
inverse problem, 94–95
inverse shuffles, 87, 94–96, 99, 101
iPhone, 23, 229
irrational numbers, 159
Isaacs, Stan, 216

Jackson, Brad, 45, 58–59
James, Stewart: background of, 182; Fibo-
nacci numbers and, 187–89; Gardner and,
187; hatred of mathematics, 181; indexes
of, 182; Jordan and, 200; Miraskill and,
184–86; modulo arithmetic and, 187–89;
number seven and, 187–89; probability
and, 185; writings of, 182–84
Jay, Ricky, xi, 119, 137, 158
Jillette, Penn, 137
Jinx magazine, 173, 175, 184, 217
*John Cage Legacy, The: Chance in Music and
Mathematics* program, 123
Johnson, Robert, 43–45
Joplin, Scott, 208
Jordan, Charles Ronlett, 190
Jordan, Charles Thornton, x, 29, 157; background of, 190–91; as chicken farmer,
190–91; Coluria trick and, 25, 54–55, 190;
de Bruijn sequences and, 189–90; dovetail
shuffle and, 193, 199–200; mind-reading

Jordan, Charles Thornton (*continued*)
and, 54–55, 190, 200; Psychola and, 193;
puzzle contests and, 191–93, 194f–198f;
radios of, 190–91; rebuses and, 191–93;
secrets of, 200
Jordan Curve Theorem, 167
"Jordan-Schönflies Theorem and the Classification of Surfaces" (Thomassen), 167
Josephus, Flavius, 98
Judah, Stewart, 166, 168–69
juggling, x–xi; ball arc and, 144–51; bouncing and, 144; combinatorics and, 138–44;
Eulerian number identities and, 143; five-ball cascade and, 143, 229; getting started
in, 145–52; going further in, 223; gravity
and, 144; handing and, 145–46; history of,
137; learning more about, 223; left-hand
to right-hand throw and, 145–46; Mills
Mess and, 144; modulo arithmetic and,
141–42; multiplex patterns and, 143–44;
parity and, 139; passing patterns and, 143;
pattern period and, 140–43; permutations
and, 138–44; Poincaré series and, 143;
scaling and, 144; sharing of tricks and, 229;
siteswaps and, 138–44; skill needed for,
137; Stirling number identities and, 143;
three-ball cascade and, 137, 139, 144–52;
three-ball high stack and, 229; tradition of
sharing tricks in, 229; Weyl group and, 143
Juggling Information Service, 144
Jürgens, H., 77

Kalush, Bill, xi, 113
Kantor, William, 93
Kaplansky, I., 99, 222
Kearn, Vickie, xi
keys, 59–60
Klondike shuffle. *See* milk shuffle
Knife through Tongue, 116
Knuth, Donald, 25–26, 80, 126, 215–16
Koran, Al, 206
Kraut, Manny, 103–4
Kruskal, Martin, 212

Laplace, Pierre-Simon, 42
Larsen, William, 25, 55
Last Recreations, The (Gardner), 214
"leave them or turn them" shuffle, 10–11
Lei, T., 77
Leibniz, Gottfried Wilhelm, 226
Leipzig, Nate, 173
Le Paul, Paul, 202
Lévy, Paul, 101–2
Liber Abaci (Fibonacci), 113–14, 187

Lie Speller, 217
linear shift-register, 28
Linking Ring magazine, 82
Little Duke playing cards, 173
Lorayne, Harry, 206–7
lying trick, 63–65

McKay, Brendan, 42–43
McMullen, Curt, 78
magic: appreciation of, 230; audiences and, 1,
3–4 (*see also* audiences); going further in,
222–23; misdirection and, 23, 223; number
of tricks and, 114–19; patter and, 11, 22,
47–48, 52, 56, 64, 71, 105, 108, 109, 128,
164, 166, 184, 218; secrets and, 92, 104,
223, 225–30
Magic Book, The (Fulves), 92
Magic Circle, The, 225
magic clubs, 153
Magic Emporium, 103
Magic magazine, 193
magic squares, 216
Magic Tricks, Card Shuffling, and Dynamic Computer Memories (Morris), 94
Magister Ludi (The Glass Bead Game) (Hesse), 125
Magnetic Colors, 82
Mandelbrot set: defining, 72; full, 76;
Gilbreath Principle and, ix, 61, 72–78;
iterations and, 72–78; leafy quality of, 72;
Möbius function and, 78; modulo arithmetic and, 80–81; online programs for, 72;
periodic points and, 74–75; shuffles and,
61, 72–78; squaring and adding, 72–77;
two dimensions, 72, 76
Mandelbrot Set, The: Theme and Variations
(Lei), 77
Markov chains, 41–42
Marks, Leo, 35–36
Marlo, Edward, 82, 91–92, 154–55
Martin, M. A., 23
Martin Gardner Presents (Gardner), 217
*Mathemagic: Magic, Puzzles and Games with
Numbers* (Heath), 216
Mathematical Association of America, 221
Mathematical Circus (Gardner), 187
mathematics: Archimedes and, 226; Banff
International Research Station and, 42–44;
calculus, 32, 118, 215, 222, 226; capture/
recapture estimation, 115; Cardano and,
226; cryptography, 30, 34–38, 211–12, 227;
de Bruijn sequences, 17–26, 30–43, 47–52,
107, 189; factorials, 6, 49–50, 68; Fermat
and, 125, 226; Fibonacci and, 113–14,

187–89; Gardner's influence in, 211,
214–21; geometry, 33 (*see also* geometry);
Gilbreath Principle, ix, 61–62, 64, 68–72,
79–83; going further in, 220–24; Gowers
and, 226–27; Hamiltonian cycles, 43–46,
227–28; Internet collaboration and, 227;
irrational numbers, 159; James and, 181;
juggling and, 137–52; Leibniz and, 226;
Mandelbrot set, ix, 61, 72–8; modulo
arithmetic, 27, 34–36, 54, 69–72, 80–81,
95, 110–11, 141–42, 187–89; Newton and,
226; NP-hard problems, 227–28; number
seven properties, 187–89; parity, 14, 94,
139, 185, 208; Pascal and, 125; Poincaré
conjecture, 221; Polymath project and,
226–27; popular view of, 1–2; practical
uses of, 30–46; prime numbers, 57–58,
78, 189, 216; randomness, 2–8, 11, 14, 18,
32, 34–35 (*see also* randomness); "real,"
xi; recreational, 211, 215, 217, 219–21;
Riemann hypothesis, 221; Royal Hummer
and, 12–16; secrets and, 225–30; siteswaps
and, 138–44; standard error, 116; Tartaglia
and, 226; topology, 162, 164–73; universal
cycles, 47–60; Wiles and, 226
Mathematics, Magic and Mystery (Gardner),
177, 217
Mathematics: The Man-Made Universe (Stein),
222
Matsuyama, Mitsunobu, xi
Matters Mathematical (Herstein and Kaplansky), 99, 222
Mazur, Barry, xi, 222
measurement error, 159
memory experts, 206–7
mentalists, 206
Method, the, 226
mice problem, 117–18
Microwriter, 208
middle-layer problem, 43–45
Miles of Tiles (Radin), 79
milk shuffle, x, 85, 96–102, 159–60
Miller, Charles, 159
Mills, Steve, 144
Mills Mess, 144
Milnor, John, 76
Minch, Stephen, 159
mind-reading tricks: Annemann and, 173; computers and, 85–92; effect and, 52–55; Elmsley and, 90; Jordan and, 54–55, 190, 200;
modulo arithmetic and, 54; Monge shuffles
and, 87–90; perfect shuffles and, 85–92
Miracle Divination: coding for, 106; effect
of, 106–12; four-object, 107–12; Hilliard

and, 105; modulo arithmetic and, 110–11; performance of, 105–6; permutations and, 107, 112; Ron's $1.98 trick, 108–12

Miraskill, 184–86

misdirection, 23, 223

Möbius function, 78

modulo arithmetic: card notation and, 27; corruption term and, 36; cryptography and, 34–36; de Bruijn sequences and, 36; Gilbreath Principle and, 69–72; James and, 187–89; juggling and, 141–42; Mandelbrot set and, 80–81; mind-reading effect and, 54; Miracle Divination and, 110–11; number seven trick and, 187–89; perfect shuffles and, 95

Monge, Gaspard, 96

Monge shuffle, x, 85, 87–90, 96–101

Moreno, Eduardo, 43

Morris, S. Brent, 94

Moschen, Michael, 144

Mosteller, Fred, 212

Mulcahy, Colm, xi

Muller, Reinhard, 83

multiplex patterns, 143–44

Music Committee of the Merce Cunningham Dance Company, 123

music files, 34

Naor, Moni, 227–28

National Security Agency, 38

Naval Postgraduate School, 43

Neale, Bob, xi; anthropology and, 161; background of, 160; End Game and, 210; Inside-Outside and, 164–73; as minister, 160; origami and, 160–64; as professor, 160; psychiatry and, 160–62; Rock, Paper, Scissors and, 162, 210–11; theatrical bent of, 161; topology and, 162, 164–73, 167

Netz, Reviel, 226

New Mathematical Diversions from "Scientific American" (Gardner), 82

Newton, Isaac, 226

New York Times, 216

Noel, William, 226

North Carolina State University, 43

Notices of the American Mathematical Society, 167

NP-hard problems, 227–28

number theory: going further on, 222; James and, 189; Möbius function and, 78; number seven and, 187–89

"order matters" study, 47–52

origami, 160–64

Out of This World, 65

out-shuffles, 92–95, 99–101, 156

over/under shuffle. See Monge shuffle

Pallbearers Review journal, 210

Paper Magic (Harbin), 162

Paper Money Folding (Neale), 162

parity, 14, 94, 139, 185, 208

Pascal, Blaise, 125

passing patterns, 143

Paul Curry Presents (Curry), 65

Peitgen, H. O., 77

Penelope's Principle, 159–60

Penrose tiles, ix, 79, 211

Penrose Tiles to Trapdoor Ciphers (Gardner), 79

perfect maps, 32

perfect shuffles: binary numbers and, 156; difficulty of, 84–85; Elmsley and, 92–95, 156; in-shuffle, 92–95, 101, 156; mind-reading trick and, 85–92; out-shuffle, 92–95, 99–101, 156; reverse, 85–88, 91–92

perfect squares, 78

periodic juggling patterns, 140–43

periodic points, 74–75

permutations: Gilbreath Principle and, 68–71, 75–80; Hamiltonian cycles and, 43–46, 227–28; hexagrams and, 121–27, 134; juggling and, 138–44; Miracle Divination and, 107, 112; Möbius function and, 78; "order matters" study and, 47–52; perfect shuffles and, 94; Royal Hummer and, 14; universal cycles and, 50–52, 55

philosophy, 30, 40–42, 122, 161, 211, 215, 223

Pilliner, Evelyn, 173

Pisano, Leonardo (Fibonacci), 113–14

Poincaré conjecture, 221

Poincaré series, 143

poker, 8, 14, 57, 64, 66–67, 93–94, 175

Polymath project, 226–27

Practical Mental Effects (Annemann), 173

Premier Partie des Subtiles et Plaisantes Invention, La (Prevost), 106, 112–16

Prevost, J., 106, 112–16

prime numbers, 57–58, 78, 189, 216

Princeton University, 101, 212, 217

probability: capture/recapture estimation and, 115; exchangeability and, 41–42; Fermat and, 125; Hamiltonian cycles and, 43–46, 227–28; history of, 125–26; I Ching and, 119–20, 125–27, 134, 136; James and, 185; Lévy and, 101–2; Pascal and, 125; randomness and, 2–8 (see also randomness); roulette system and, 179–81; shuffles and, 81; standard error and, 116; uncertainty and, 125

Problems Plaisants et Delectables Qui se sont par les Nombres (Bachet), 107–8, 115–16

products, 52

protein folding, 42

psychiatry, 160–62

Psychola, 193

psychological force, 158

Puzzles Old and New (Slocum and Botermans), 218

quasicrystals, ix, 79

Quasicrystals and Geometry (Senechal), 79

Quinn, Frank, 219

Radin, Charles, 79

Ralston, Anthony, 25

Ramnath, S., 95

Rand Corporation, 82

randomness, 2; code making and, 34–35; computers and, 48; de Bruijn sequences and, 18; factorials and, 6, 68; Gardner and, 215, 218; genetics and, 41–42; Gilbreath Principle and, 61, 63, 66; Gray Codes and, 25–26, 42–43; Hamiltonian cycles and, 228; Hummer and, 204–5, 209; I Ching and, 120, 122, 124–27, 133–36; James and, 185, 188; Jordan and, 193, 200; mind-reading computer and, 89–90; Neale and, 160, 164; number of magic tricks and, 115–16; robotic vision and, 32; Royal Hummer and, 8; "turn two and cut at random" and, 4–7, 11, 14; universal cycles and, 47–49, 59

"Reading the Fifty-two Cards after a Genuine Shuffle" (Wiliams), 193

rebuses, 191–93

"Récréations Mathèmatiques (1624): A Study on Its Authorship, Sources, and Influence" (Heeffer), 113

red/black trick, 61–66

Reed College, 58

Reeds, Jim, 200

Riemann hypothesis, 221

"Riffle Shuffle Card Trick and Its Relation to Quasicrystal Theory, A" (de Bruijn), 79

riffle shuffle, 84; Elmsley and, 157; Gilbreath Principle and, 62–70; Jordan and, 193, 199–200; Mandelbrot set and, 61, 79

Riser, Harry, 191

Rivest, Ron, 38

Robbins, Herbert, 222

robotic vision, 17, 30–34

Rock, Paper, Scissors, 162; Endfield and, 208–10; Hummer and, 210–11; James and, 185–86; Neale and, 210–11

Rogers, T. D., 77
roulette system, 177–81
royal flush, 8–15
Royal Hummer, 8–16
Royal Road to Card Magic (Hugard and Braue), 222
RSA, 38
Rusduck, 93–94
Ruskey, Frank, 26, 43

Sands of the Kalahari (film), 208
San Jose State University, 58
Saupe, D., 77
Savage, Carla, 43
Say Red, 181
S-boxes, 37–38
Schwab, Richard, 118
Science of Conjecture, The: Evidence and Probability before Pascal (Franklin), 125
Scientific American magazine, 82, 211, 214–15, 217
Scot, Reginald, 106–7, 113–16
Scully, D., 95
"Secret Gimmicks" (Jordan), 200
secrets, 7–8, 15, 158, 202; Christ and, 173; codes and, 17 (*see also* codes); de Bruijn sequences and, 18–24; Gilbreath Principle and, 63–65; Inside-Outside and, 164–73; James and, 182; Jordan and, 190–91, 200; Miracle Divination and, 105; nature of magic and, 92, 104, 223, 225–30; Rock, Paper, Scissors and, 209–10; sleight of hand and, 154
"Secrets of the *I Ching* Revealed" (Diaconis), 119–20
Senechal, Marjorie, 79
seven, properties of, 187–89
Seber, George, 116
Shamir, Adi, 38
Shannon, Claude, 200
shift registers, 28, 49
shuffles: Australian (down-and-under), x, 85, 88–89, 91, 98–102; CATO, 7; combining, 93; cycle structures and, 78, 102; dovetail, 193, 199–200; false, 159; faro, x, 98; Gilbreath Principle and, 61–72, 79–83, 211; in, 92–95, 101, 156; inverse, 87, 94–96, 99, 101; "leave them or turn them," 10–11; Mandelbrot set and, 61, 72–78; milk, x, 85, 96–102, 159–60; mind-reading trick and, 85–92; Monge, x, 85, 87–90, 96–101; out, 92–95, 99–101, 156; perfect, 84–95, 98–102, 137, 154, 156, 159–60; probability and, 81; relationship among, 99–102; riffle,

61–70, 79, 84, 157, 193, 199–200; roulette system and, 177–81; Royal Hummer, 8–15; stay-stack arrangement and, 88–90, 94, 154; "turn two and cut at random" (Hummer), 4–7, 11, 14
Sinden, Frank, 30
Si Stebbins order, 54
siteswaps, 138–44
sleight of hand, 103, 222–23; Annemann and, 173; Cardini and, 173; Christ and, 173, 177; Elmsley and, 156, 159; Findley and, 173; Gardner and, 211; Gilbreath Principle and, 66–67; Hummer and, 202, 206, 208; Judah and, 168–69; Leipzig and, 173; mind-reading computer and, 91–92; Miller and, 159; Miracle Divination and, 106; Neale and, 169, 171; Penelope's Principle and, 159–60; Vernon and, 159, 171, 173
Slocum, Jerry, 218
Smith, K., 122
Society of American Magicians (SAM), 82, 216, 222–23
solitons, 212
Solt, John, 119
"Solutions to Elmsley's Problem, The" (Diaconis and Graham), 95
"Sonata for Prepared Piano" (Cage), 122
Song Dynasty, 120
Sphinx magazine, 190
squaring and adding, 72–77
standard error, 116
Stanford University, 216, 221
Staple, Aaron, 48
stay-stack arrangement, 88–90, 94, 154
Steele, Richard, 104
Stein, Sherman K., 25, 222
Steiner point, 212–14
"Steiner Problem, The" (Gardner), 213–14
Stirling numbers, 143
Stirling's approximation, 49
Suitability, 25, 55
Sung Dynasty Uses of the "I Ching" (Smith), 122
superblock striping technique, 80
Su Wu Tending Sheep, 130–33
swindles, 4, 162
Swinford, Paul, 95

Tamariz, Juan, 229
Tannen, Louis, 103–4, 156
Tarbell Course in Magic, 191
Tartaglia, 226
Ten-Card stars, 168
Tenner, Ed, xi
This Is Not a Book (Neale), 162, 210

Thomassen, Carsten, 167
Thompson, Silvanus P., 222
three-ball cascade, 137, 139, 144–52
three-ball high stack, 229
three-initials trick, 218–19
Thurston, Howard, 104, 216
Thurston, William, 76
topology, 162; Inside-Outside and, 164–73; Jordan Curve Theorem and, 167
Tops magazine, 92
trampolines, xi
Trost, Nick, 169
"turn two and cut at random" shuffle, 4–7, 11, 14
Tversky, Amos, 63

Union Theological Seminary, 126, 160
universal cycles: Bell numbers and, 55–57; de Bruijn sequences and, 17–26, 43, 47–52; defined, 55; genetics and, 39; mind-reading effect and, 52–55; "order matters" study and, 47–52; permutations and, 50–52, 55; randomness and, 47–49
University of Victoria, 43

vanishing square, 217
Veeser, Bob, 92
Vernon, Dai, 91, 154, 158–59, 171, 173, 217, 230
Voodoo Fortune Telling, 186

Weiss, A., 77
Weyl groups, 143
What Is Mathematics? (Courant and Robbins), 222
Where's Waldo game, 227–28
Whole Art and Mystery of Modern Gaming (Anonymous), 96
Wikipedia, 222, 229
Wiles, Andrew, 226
Wiliams, C. O., 193
window shapes, 32–33
Wohl, Ronald, xi, 25, 47, 52, 54–55, 66
Wood, Sherry, xi
Woodson, George Mrs., 190
Wright, T. Page, 25, 55
Wurlitzer Building, 103

YouTube, 153, 222, 229

Zarrow, A Lifetime of Magic (Ben), 67
Zarrow, Herbert, 66–67
zero-knowledge proof, 227
Zulu (film), 208